The Technology Gauntlet

The Technology Gauntlet
Meeting the Challenges of Workplace Computing

Margaret Kilduff
New Jersey Institute of Technology
Newark, New Jersey

Doug Blewett
AT&T Bell Laboratories
Murray Hill, New Jersey

ADDISON-WESLEY PUBLISHING COMPANY
Reading, Massachusetts   Menlo Park, California   New York
Don Mills, Ontario   Wokingham, England   Amsterdam   Bonn
Sydney   Singapore   Tokyo   Madrid   San Juan
Paris   Seoul   Milan   Mexico City   Taipei

Many of the designations used by manufacturers and sellers to distinguish their products are claimed as trademarks. Where those designations appear in this book and Addison-Wesley was aware of a trademark claim, the designations have been printed in initial caps or all caps.

The publisher offers discounts on this book when ordered in quantity for special sales. For more information please contact:

Corporate & Professional Publishing Group
Addison-Wesley Publishing Company
One Jacob Way
Reading, Massachusetts 01867

Library of Congress Cataloging-in-Publication Data

Kilduff, Margaret
    The technology gauntlet : meeting the challenges of workplace computing / Margaret Kilduff and Doug Blewett.
        p.   cm
    Includes bibliographical references and index.
    ISBN 0-201-63359-0
    1. Management information systems. 2. Technology transfer.
I. Blewett, Doug, 1949-   . II. Title.
T58.6.K546   1994
658.4'038--dc20   94-18016
                                                          CIP

AT&T Copyright © 1994 by AT&T, Margaret Kilduff

ISBN 0-201-63359-0

Text printed on recycled paper
1 2 3 4 5 6 7 8 9 CRW 97969594
First Printing, September 1994

Contents

Preface

No, confound her, her intellect was good, she had brains enough, but her training made her an ass....

Mark Twain in *The Connecticut Yankee in King Arthur's Court,* Chapter XVII, In the Queen's Dungeons.

So said Hank Morgan, a nineteenth-century resident of Connecticut, of a sixth-century woman in Mark Twain's *A Connecticut Yankee in King Arthur's Court.* Much about sixth-century training annoyed Hank Morgan. One aspect was how sixth-century training made nineteenth-century technology almost incomprehensible. We assume that Mark Twain felt that some nineteenth-century training also made nineteenth-century technology incomprehensible. We know that the training some people receive in the twentieth century makes the technology of this century incomprehensible. This concerns us. Although we are concerned about the understanding and use of all current technology, we have particular concerns about the use of workplace computing, about the use of computers to handle the information needs of the workplace.

We have seen many people of good intellect brought to their knees (often along with their businesses) by inadequate training in workplace computing. Sadly, most of these people realized they were inadequately trained only in hindsight. That is the strange thing about inadequate training, you often do not realize how it is affecting you until it catches up with you. This book is designed to help you get the training you need to use workplace computing effectively. The key to the training is not specifics about computers or computing. The key to it is a method for thinking about computing (or any technology for that matter). This method (a technology transfer method) is based on our almost 20 years of living daily with computers and the associated technology. During this time we have written software, taken apart hardware (and put it back together with no pieces left over), served as local experts, and been both formal and informal consultants to individuals and businesses (of all sizes, but primarily small- and medium-sized businesses). The technology transfer method is discussed in the first eight chapters of this book. Chapter 9 presents information about choosing a consultant (for those who wish to do so). A main point of this book, however, is that you do not usually need a consultant to use workplace computing effectively. Chapters 10 and 11 present information about computing and computers, and Chapter 12 summarizes of the questions that need to be answered to complete the technology transfer method successfully.

Although the book focuses on automation technology, the method applies equally well to most new technologies (or "old" technologies that are new to you). The automation technology transfer process should be effective with reasonably sized organizations regardless of whether the organization has an information/computing services department. It should be effective regardless of whether the organization wants to automate one function or all functions. The process is also applicable to the entire range of system development possibilities from the purchase of "off-the-shelf" material to the purchase of a customized system.

We tried to design the book so that you can "pick and choose" the chapters to read without wondering too much what happened in the chapters you did not read. We tried to make the chapters relatively independent of each other. Some readers may be interested in reading about consultants (Chapter 9) and some not. Some may be interested in reading about ways to choose computer products (Chapter 7) and some not. The quotes at the beginning of each chapter are from Mark Twain's *A Connecticut Yankee in King Arthur's Court*. So much of what Twain wrote is applicable to technology transfer in any age. Indeed, you could call Twain's *Yankee* book an interesting case study of an attempted technology transfer effort that failed miserably. Perhaps, if the characters in the book had used the techniques discussed in this book, the story would have had a much different ending.

We also tell two continuing stories throughout the book. We use these stories to illustrate points about the method. One story is about Chuck and his efforts to automate his social service agency. The other story is about Carla and her efforts to recover from a computing disaster at her stevedore and warehousing company. The two stories are business case studies in the form of "docudramas." They are based on real situations, but they are by no means exact histories of either. Nor is the description of each situation excessively detailed. We tried to include enough detail to illustrate key points, but not so much detail that the stories would be entirely boring.

So who should read this book? Primarily people in small- and medium-sized businesses or organizations who are wondering how to incorporate computing into the business. Reading this book should help people avoid miserable, disappointing, and hostile computer project experiences. Anyone who has had a bad computer experience and now dislikes computers may also find this book helpful. By reading this book, such people may discover they have been misdirecting their computer hostility — the method of automation and not computers may have caused the bad experience. Members of the business community and other individuals who never intend to automate because they see no utility in automation may also find the book informative. By reading this book such people may discover that computers are relevant and potentially important to their lives.

Parts of this book should also be of interest to those who serve as consultants, those working in the computer industry, and those interested in the transfer of technology. Anyone interested in the dynamics of keeping afloat in the sea of technology that seems to be constantly washing over us should also find the book interesting, as should anyone who is managing an active, growing technology. Anyone who is not in a group mentioned above (and we tried very hard to include everyone on the planet) might want to skim the book to look at the pictures. So who should find this book interesting? Abraham Lincoln really summed up our view of this book's audience:

People who like this sort of thing will find this the sort of thing they like.

We wish to thank all those who have contributed to the development and completion of this book. There are hundreds of people who have worked beside us, encouraged us, and influenced our thinking over the years. Those deserving our special thanks are: Scott Anderson, Megan Blewett, Betty Brokow, Sally Browning, Kim Dawley, Rich Drechsler, Kathleen Duff, Roger Gourd, Kate Habib, Ken Hicks, John Odom, Simone Payment, Marty Rabinowitz, Lisa Raffaele, John Wait, and Deb Wilde. We are greatly indebted to their good advice, insights, hard work, and support.

"Perspective," digital enhancement by Thomas Micchelli of a reprint of a drawing done in 1605 by Henricus Hondius. The original reprint is from the Dover Pictorial Archive Series, "Perspective," by Jan Vredeman de Vries, Dover Publications, Inc., New York, New York, 1968.

Chapter 1
Technology's Gauntlet

....this was challenging death in the open field unarmed, with all the odds against the challenger, no reward set upon the contest, and no admiring world in silks and cloth of gold to gaze and applaud....

Mark Twain in *The Connecticut Yankee in King Arthur's Court*, Chapter XXIX, The Smallpox Hut.

Life's challenges take many forms. These challenges form gauntlets of many shapes. Some challenges last a moment. Others last a lifetime. Some gauntlets contain challenges that are easy to overcome. In other gauntlets, "all the odds are set against the challenger." Successfully running some gauntlets brings great rewards. For others, there is "no reward set upon the contest." The challenges in some gauntlets are intentional and well defined. Consider a medieval jousting tournament or the phrase "Make my day." The challenges in other gauntlets are unintentional and ill defined. Such were the challenges that Hank Morgan (a nineteenth-century Connecticut Yankee) created for the members of King Arthur's court. Such were the challenges that the Court, in turn, created for the Yankee. Each created an unintentional gauntlet of unintentional challenges for the other to run. The gauntlets created by the Court and the Yankee for each other were unintentional because neither meant the other any harm. Each meant only to use the world paradigms they had always used — to view and explain the world as they always had. Their respective paradigms (e.g., about government, nature, health) were, however, often incompatible. Yet this incompatibility had little impact on their respective paradigms. Each persisted — privately and often publicly — in the correctness of their views. Nor could the incompatible paradigms and the people who held them be ignored. They were all in the same place at the same time with no place else to go. The Yankee and the Court were bound together in all their incompatibility. This created enormous daily unintentional challenges for each. These unintentional challenges that each created for the other formed unintentional gauntlets that each had to run; gauntlets in which each often felt trapped with no hope of a successful run.

The Yankee and the Court are clearly not the only fictional or nonfictional people to be unintentionally challenged by a new incompatible and persistent paradigm. Consider, for example, the attitude and feelings of seventeenth-century Vatican officials toward Galileo's solar system ideas. And it is not just new paradigms that created unintended challenges. An "old," alternate, incompatible, and persistent paradigm can be as challenging as a new one [1.1]. Consider the number of people past and present who feel challenged by Darwin's theory of evolution. These challenges lie in the area where two incompatible paradigms are forced to overlap — forced to co-exist (see Figure 1.1). The overlap can sometimes be tolerated for a brief time. If the incompatible paradigms are persistent, however, the challenges can become quite wearing. Often a new paradigm quickly creates new technologies (new techniques, new ways of doing things, new tools, new machines, new artifacts). Atomic physics, for example, led to atomic weapons. New technologies also result from the logical extension of "old" paradigms. Electromagnetic theory led to microwave communication towers and microwave ovens. And still other new technologies are "merely" the most recent evolution or major improvement on "old" technologies. Improvements in plastics and plastics engineering led to the plastic, nonstick snow shovel; an improvement dwarfed only by the bent — instead of straight — snow shovel handle (as anyone who has ever shoveled wet snow clearly knows).

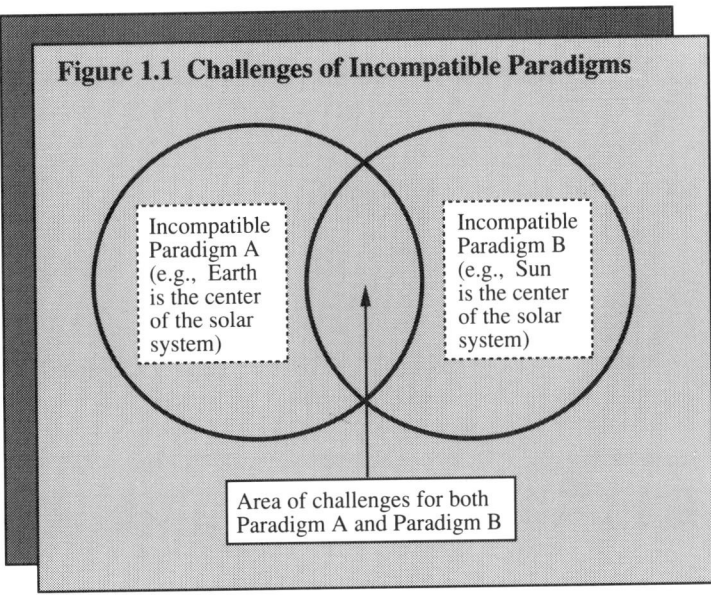

Figure 1.1 Challenges of Incompatible Paradigms

Regardless of origin, however, a new technology (often referred to today as "high-tech") frequently creates an enormous unintentional gauntlet that must be run by individuals, businesses, communities, and society. This gauntlet exists when the technology is both persistent (unable to be ignored) and incompatible (in whole or in part) with current daily life. No technology better illustrates this high-tech gauntlet than workplace computing — the use of computers to help a business meet its information-related needs (e.g., maintaining, accessing, and organizing company records; preparing budgets; preparing schedules; preparing reports; running payroll; sending personalized letters; generating accounts receivable forms). The unintentional challenges of workplace computing (and other types of high-tech) usually take the following forms:

- The challenge of deciding how much, if any, of the technology to incorporate into daily life.

- The challenge of productively incorporating the chosen technology into daily life (if the decision has been made to do so).

- The challenge of keeping up with changes in the technology.

- Sometimes, unfortunately, the challenge of dealing with the technology when it runs amok.

Throughout this book we will follow two companies as they meet these four challenges in terms of workplace computing. The first two challenges are faced by a social service organization headed by Chuck. A warehousing and stevedore company headed by Carla faces the latter two. In this chapter we introduce the two companies and their problems.

Chuck's Social Service Organization

Chuck runs a nonprofit social service organization employing 31 people (5 are part-time). Half of his annual budget of $800,000 comes from seven different funding sources (e.g., United Way, various foundations, government, private contributions) and the other half from client fees. Each year, Chuck submits funding requests for the next year to a variety of funding sources. He also submits accountability reports to the current funding sources. Others (e.g., the Board of Trustees) also receive periodic or annual accountability reports. The completion of the funding requests and accountability reports is becoming increasingly difficult and time-consuming. Each year, more and more detailed information is required (e.g., the number and percentage of clients who are children, the percentage of administrative costs paid from institutional sources, the effectiveness of services, budget projections). Chuck does not have such information easily available. To complete the requests and reports, Chuck sorts through index cards, notebooks, log sheets, and paper files. He writes the requests and reports longhand and then a secretary types them. Chuck is increasingly concerned that he will miss the request and report deadlines and that he will not be able to provide all the required detailed information. Failing to meet the deadlines or provide the information decreases the probability that Chuck will obtain requested outside funding. The outside funding is essential funding — without it Chuck's social service organization cannot continue to operate (see Figure 1.2).

Chuck has recently come to believe, however, that if he automates all of the organization's paper records, his request and reporting problems will be solved. Chuck believes that with the use of workplace computing all necessary information will be easily and quickly available; that the submitted material will look more professional (e.g., elaborate charts and graphs, no typos or obvious corrections, better print quality); and that he will be able to produce more extensive and detailed statistical analyses. He believes this partly because others have told him it is true, partly because of what he has seen computers do in other organizations (e.g., in Chuck's bank, physician's office, supermarket), and partly because off-site automated processing of the organization's payroll has been very effective. (Chuck has been outsourcing payroll production for years.) When payroll production was automated through an outside vendor, the entire payroll production process began to work more smoothly. Chuck used to think that workplace computing (other than for payroll) was fairly irrelevant to his organization. Chuck now sees extensive use of workplace computing as essential just to "keep up."

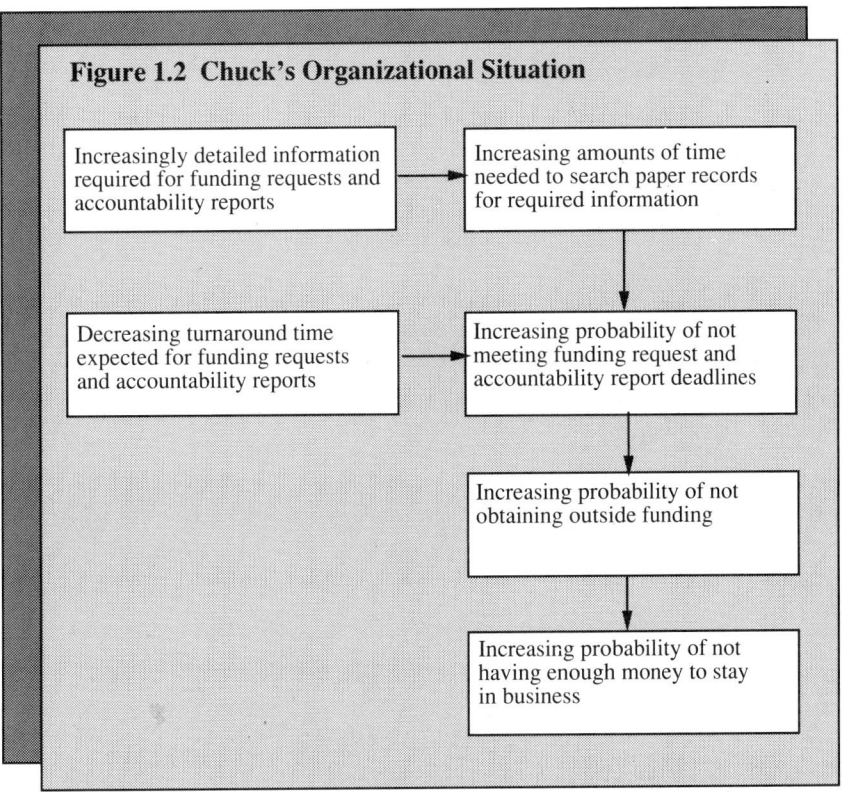

Figure 1.2 Chuck's Organizational Situation

At the same time that Chuck sees extensive workplace computing as absolutely essential to the organization's ability to survive, he does not know how compatible it is with his organization. He does not view either himself or other agency personnel as being particularly technical; he fears that computing is incompatible with the current skills of employees (including himself). How would they ever learn to use workplace computing productively? Chuck, in addition, has little idea about what is available and applicable in terms of workplace computing. Nor does he know how to tell "good" workplace computing from "bad" workplace computing. He does not know how much workplace computing he needs. Chuck also has absolutely no idea how he can afford to automate; computing seems incompatible with the organization's budget. Some of the specific challenges in Chuck's gauntlet are shown in Figure 1.3.

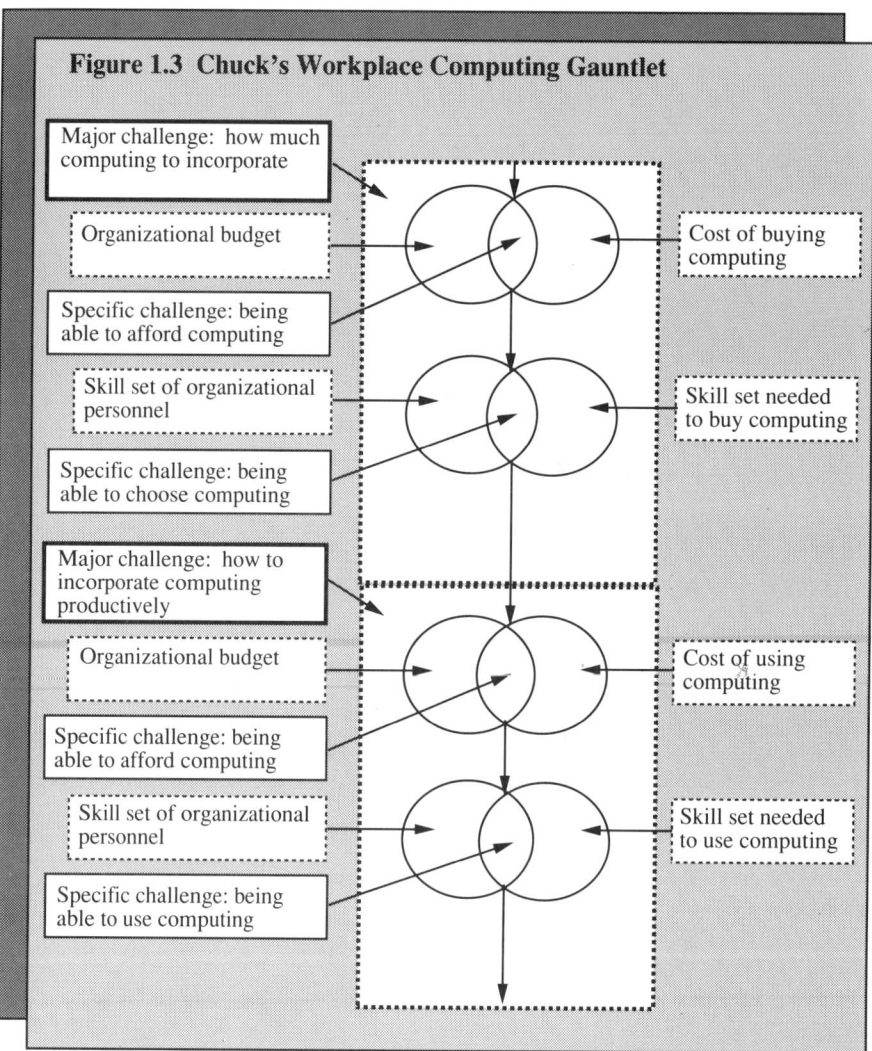

Figure 1.3 Chuck's Workplace Computing Gauntlet

These specific workplace computing challenges can, however, be organized into two major categories or areas:

- The challenge of deciding how much, if any, workplace computing to incorporate into the daily life of the social service organization.

- The challenge of productively incorporating any chosen workplace computing into daily life (if the decision is made to do so).

Chuck's situation also illustrates a number of additional points about organizations and technology.

- Technology is often unevenly distributed.

- Technology often changes daily standards.

- Today's businesses (especially small- to medium-sized businesses) often feel trapped by technology.

Despite the fact that workplace computing seems to be everywhere, Chuck and the members of his organization really have little hands-on experience with workplace computing. For them, workplace computing is as much of an unknown — as much of a frontier — as it was for the first few users of workplace computing. This is but one illustration of the uneven distribution (or diffusion) of technology. A particular technology can be "everywhere" in some countries, industries, companies, departments, communities, or schools and nowhere to be found in others [1.2–1.8]. Not all lodging, for example, in all countries has indoor plumbing or electricity. Not all people in all countries have easy access to immunization shots. The uneven diffusion of a technology creates a wide range of familiarity and proficiencies with that technology — from "never heard of it" to "use it all the time and cannot live without it." Proficiency levels also vary even when fairly easy access to a technology exists. Consider, for example, the wide range of driving proficiencies found on any highway in the United States. And each year new drivers (primarily adolescents) must be trained. The existence of varying proficiency levels is somewhat obvious, yet it seems to us that this situation is often ignored by the people most interested in seeing a particular technology used (e.g., the researchers, developers, sales personnel, and trainers working with that technology). We have known people who seem to believe that once some critical mass of people is proficient with a technology, proficiency travels almost telepathically throughout the country, industry, company, department, community, or school. Such people seem to believe that the hard work done to develop proficiency in one person does not have to be repeated with the next. This is not true. Proficiency with any technology is developed one person at a time. It is developed by helping each person negotiate his or her own personal technology gauntlet.

Chuck's situation also illustrates the relationship between a persistent technology and daily standards. A persistent technology often becomes more prevalent; an increasing number of people use it. As this happens, the technology changes the way things are done. The technology usually creates new standards for daily life [1.9–1.12]. Houses without indoor plumbing, for example, were once the norm in the United States. Now such houses are considered substandard. One of the reasons that Chuck is feeling increasing information pressures is because the persistence of computing has changed information daily standards. Chuck can no longer keep up with reporting requirements because computing has changed the standards for reporting. It was once the norm for Chuck (and others) to take at least one week to produce a funding request. Numerical calculations were made using a calculator and the request was typed on an electric typewriter. Now taking one week to produce the request is a substandard process. Others in Chuck's industry can produce the funding request in less than one week. Chuck must too if his organization is to compete successfully for funding. The automation requirements of any one company within an industry, therefore, depend as much on the automation used by others in the industry as the company's own goals. If only one company in an industry automates, then the industry as a whole has entered the automation era. If automation of that one company gives the company any advantage, then others in the industry are usually forced to automate to remain competitive.

Clearly, once a technology changes the daily standards, nonuse of the technology becomes difficult, if not risky. Yet use of the technology creating the standard can also be difficult. It can be incompatible (in whole or in part) with current daily life. It can be incompatible with the budget and other organizational resources. This is the situation in which Chuck finds himself. He is caught between a rock and a hard place; he feels trapped by the technology of workplace computing. Chuck does not really view the existence of workplace computing as a wonderful opportunity for him. He views its existence as another way that he and his organization can get behind. Chuck views workplace computing as another demand on already scarce organizational resources. Chuck is not alone. Many small- to medium-sized businesses feel more demoralized than excited by the existence of various technologies. Such businesses see the technologies and changing daily standards around them; the businesses need and want to meet the standards, but do not have the organizational resources to do so. Small- to medium-sized businesses rarely have extra resources. As technology becomes more prevalent, the cost of doing business increases. It used to be, for example, that you could probably start a business if you had expertise in one good idea plus a desk, chair, telephone, paper, and typewriter. Now you need at least one good idea plus a desk, chair, telephone, paper, voice-mail, fax machine, computer, and expertise in using all this technology. You also need more highly trained personnel who can use all the high-tech equipment [1.13].

Carla's Stevedore and Warehousing Company

Carla runs her family's 60-year-old stevedore and warehousing company. The company has been using workplace computing for years to produce paperwork for the warehousing operations (e.g., invoices, inventory lists) and payroll for the company's 2500 employees (primarily stevedores and warehouse personnel). Warehousing paperwork is generated daily and the payroll is run weekly. All the computer work is handled by the company's Computing Services Department (CSD) located 10 miles from the warehouses. CSD employs 10 full-time people (primarily operations personnel, data entry people, programmers, and systems analysts). Eighteen months ago the company purchased a new mainframe computer. This computer has been a nightmare almost since the day it arrived; the machine has been down (not available for production work) more than it has been up (available for production work). Four days ago — on Monday — the machine seriously malfunctioned and almost put Carla and her company out of business.

Many times during the last few days it seemed that the mainframe would never run properly (if at all) again. This created an enormous problem for Carla because it is impossible for Carla to generate necessary business information without the computer. Without this mainframe, there is no warehousing paperwork — making it difficult if not impossible to get customers' inventory in and out of the warehouses. More critical, however, is the lack of a payroll. Carla's contract with the stevedores clearly states that paychecks must be on the docks ready to distribute by 9 a.m. each Thursday. If the paychecks are not there at 9 a.m., the stevedores stop working, form a line, and wait for the checks; they wait on the clock. Every hour they wait costs Carla almost $100,000. If they wait more than half a day, it becomes likely that, in addition, all customers scheduled for that day will sue Carla. Goods must move off the ships so that other goods can move on. The ships have deadlines of their own to meet. If the machine had not been fixed, Carla's company would have been nearly bankrupt by the end of the week.

On Monday evening, the company that sold Carla the machine sent eleven field engineers, technicians, and supervisors to fix the machine. This crew knew this site and the CSD staff well since most of them had been trying to fix this machine's problem on and off for the last 18 months. The last few days had been a strenuous cycle of hope and despair; the machine would function well for a brief time and then fail. There was even serious talk of flying most of the CSD staff to another site 800 hundred miles away to run the payroll. This site was one arranged at the last minute. It was not a contracted back-up site so even if they had gone, there was no guarantee that the business software would successfully run there or that CSD would be given enough time to complete the payroll by Thursday morning. At the last possible minute, the machine began to function well enough to run the payroll. The paychecks were only half an hour late. Carla was warned, however, that some action regarding the machine needs to be taken immediately. The software needs attention too. It was written in house years ago with all updates and changes being kludged in (patchworked in; something like continuing to fill in the potholes instead of just repaving the road entirely). Because of both the hardware and software problems, it is only a matter a time (perhaps a brief time) before it will be absolutely impossible to do any work with this mainframe (see Figure 1.4).

The 18 months of problems have wreaked havoc with the machine's schedule. The preferred schedule is to run the machine continuously from 7 a.m. Monday until 11 p.m. Friday. Preventive maintenance is to be done, as needed, on the weekend. With all the setbacks, however, the machine has been running production work at least some of the weekend for more than a year. The last few months the machine has been available for preventive maintenance for only 10 hours each week. This schedule has been hard on the machine, but it has been harder on the CSD personnel. Working at CSD has ceased to be a job, it has become an endurance test.

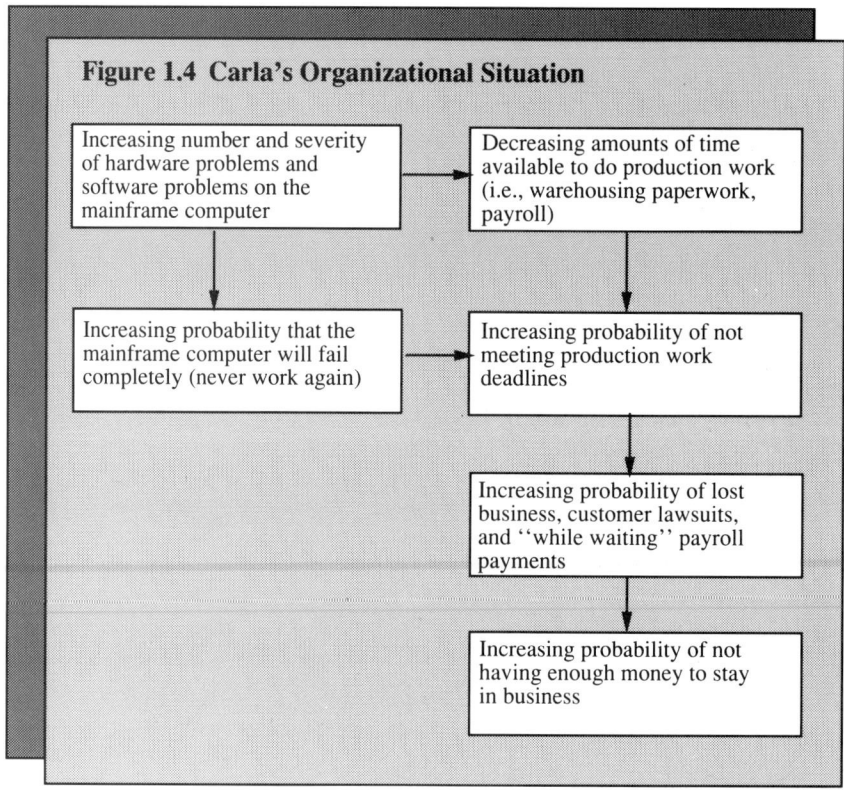

Figure 1.4 Carla's Organizational Situation

Everyone blames everyone else for the current problems. The salesman who sold Carla the machine was asked by his company to return his sales commission. He quit instead. Half the people involved in trying to keep this machine running have gained 20 pounds each over the last 18 months. The other half gained 30. All have come to view antacids as a source of nutrition. Some of the specific challenges in Carla's gauntlet are shown in Figure 1.5. These specific workplace computing challenges can, however, be organized into two major categories or areas:

- The challenge of keeping up with changes in workplace computing.
- The challenge of dealing with workplace computing run amok.

Keeping up with workplace computing changes means not only being aware of what is happening in the workplace computing industry (e.g., what are the new products, are they any good, are they relevant to the organization), but also keeping up with changes in the organization's current computing equipment (hardware and software). Some of these latter changes are made intentionally: equipment is added, functionality is added, upgrades are made.

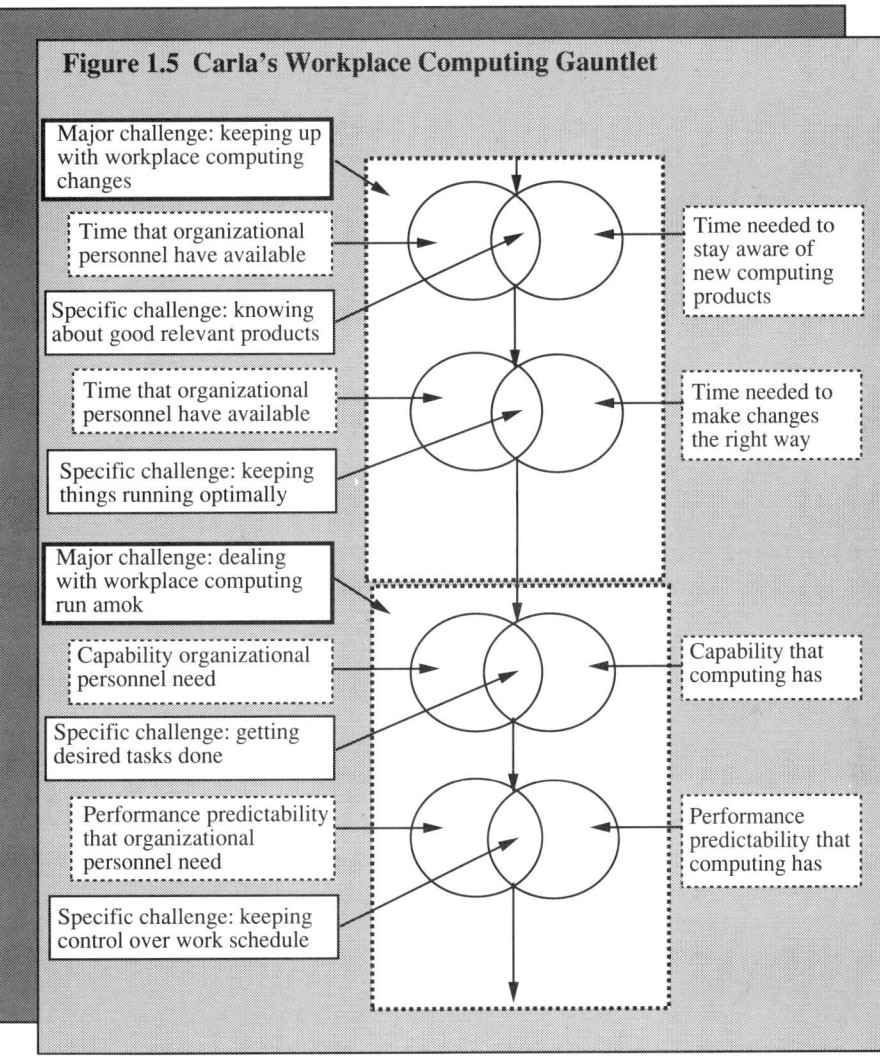

Figure 1.5 Carla's Workplace Computing Gauntlet

Major challenge: keeping up with workplace computing changes

Time that organizational personnel have available

Specific challenge: knowing about good relevant products

Time that organizational personnel have available

Specific challenge: keeping things running optimally

Major challenge: dealing with workplace computing run amok

Capability organizational personnel need

Specific challenge: getting desired tasks done

Performance predictability that organizational personnel need

Specific challenge: keeping control over work schedule

Time needed to stay aware of new computing products

Time needed to make changes the right way

Capability that computing has

Performance predictability that computing has

Keeping up with such intentional changes means making the changes in a way that a smoothly functioning system is maintained. Sometimes (as in Carla's case), most of these changes are kludged in until the system becomes unwieldy. Other changes are unintentional and inevitable. These are the basic changes caused by the regular use of the equipment — the normal wear-and-tear changes. Keeping up with these changes means regular maintenance. Carla's computing system is not being properly maintained. There is simply no time to do so. When workplace computing runs amok, it seems to have a life of its own. It (and not you) seems to be in control. It does not do what you expect it to do. Workplace computing can run amok because the equipment is broken or flawed. It can also run amok if it is being misused, if its capabilities are misunderstood. Carla — although she does not know it yet — overestimates what her computing equipment can do. It will never meet her expectations.

Carla may not know why her computing situation is a disaster, but she does know that she is very disappointed and confused by this situation (everything used to work so well). She firmly believes in the advantages of using automation. It never occurred to her that there might ever be disadvantages. Carla initially saw automation as a way to increase her company's productivity and give her company a competitive edge. Now it is close to driving her out of business. How could this have happened? The specifics of why Carla misunderstands the workplace computing equipment (and what needs to be done) are discussed throughout this book. But Carla is not the only person who has ever done this — she is certainly not the only person whose workplace computing has run amok. Nor is Carla's company the only one that has ever been disappointed with automation results. Disappointed companies usually find that workplace automation did not solve existing workplace problems and may even have created new ones. Computer use did not produce expected changes. This is certainly a major cause of Carla's disappointment. Organizations often expect computer use to produce a dramatic increase in productivity, cost effectiveness, product quality, or some combination of the three. Workplace life should improve. Sometimes, however, computer use creates minimal, or even negative, changes in productivity, cost effectiveness, or product quality [1.14–1.17]. Expectations are not met. The company is disappointed; workplace life often becomes miserable. If automation disappointment becomes automation disaster, the company might go out of business.

When expected changes do not occur (and the company survives), the automated system is usually abandoned in whole or in part or redesigned and rewritten (sometimes extensively and endlessly). Automation is more likely to produce the expected changes when — as we discuss in later chapters — automation is viewed as a tool to enable business processes. Sometimes, unfortunately, automation is viewed as the business process itself; just having automation is an improvement. Automating a nonoptimal business process does not improve the process. The familiar phrase of "garbage in, garbage out" (GIGO) applies to the automation of business processes. Automating a bad or ineffective business process does not improve company functioning. If the business process is not optimal, a company needs business process reengineering not "just" automation [1.18–1.20]. Automation is just one business process reengineering tool.

How to Meet the Challenge

Chuck and Carla both see their current computing challenges as massive, almost overwhelming. And in many ways they are. How can Chuck and Carla meet these challenges? What should Chuck and Carla do? They have many choices; life's challenges can, after all, be met in many ways. Chuck and Carla can choose to meet their challenges only partially or halfway; some or most of the challenge can be ignored. Chuck and Carla can also meet their challenges "in the open field unarmed," unprepared, or underprepared. These methods are, however, either heroic or foolhardy. It is better to be prepared. Developing adequate preparation to meet massive overwhelming challenges is difficult (to say the least), but becomes easier if the massive challenge can be sub-divided into a series of smaller ones. A massive, overwhelming gauntlet should be run as smaller, more doable gauntlets. This is the way Chuck and Carla will meet their massive computing challenges. They will subdivide their overwhelming gauntlet into six smaller ones (six decision stages) using the process described in this book. Within each of the six decision stages are questions that, when answered, provide information necessary to negotiate the gauntlet successfully. The described process is a technology transfer process; the basic challenge facing Chuck and Carla is a technology transfer one.

We define successful technology transfer as knowing what you need to know about a technology to use it productively in your life. Chuck and Carla (and many of the people who work in each of their companies) simply do not know what they need to know. Some do not know enough; some have misinformation. The successful transfer of workplace computing technology requires more than just placing a computer on a person's desk. That person needs to know how to use that computer productively; to understand exactly how the computer fits into his or her life. What happens when a computer is merely placed on a desk without the computing technology being transferred? The most common result that we have seen is that the computer becomes a large desk ornament instead of a productive work tool. The computer is rarely used. The person does not know how or why to use the computer. The described technology transfer process provides a structure for thinking productively and critically about a technology. It provides a structure to help you decide in what way, if any, a particular technology fits into your life. If it does fit, then the process provides a way to incorporate the technology. Note that successful technology transfer involves not only deciding when and how to use a technology, but also when and how not to use it. Consider antibiotics; consider nuclear weapons.

References

1.1: Kuhn, Thomas S., *The Structure of Scientific Revolutions,* 2nd ed., University of Chicago Press, Chicago, Illinois, 1970.

1.2: Tilton, John E., "International Diffusion of Technology: The Case of Semiconductors," The Brookings Institution, Washington, DC, 1971.

1.3: Mann, Charles K., and Ruth, Stephen R., *Expert Systems in Developing Countries: Practice and Promise,* Westview Press Inc., Boulder, Colorado, 1992.

1.4: Rogers, Everett M., *Diffusion of Innovations,* 3rd ed., The Free Press, New York, 1983.

1.5: Kantrowitz, Barbara, Chideya, Farai, and Biddle, Nina Archer, "The Information Gap," *Newsweek,* March 21, 1994, page 78.

1.6: Lidtke, Doris K., "Computers in Schools: Past, Present, and How We Can Change the Future," *Communications of the Association of Computing Machinery,* Volume 36, Number 5, May 1993, pages 84–87.

1.7: "What Role for Technology? Is Technology Part of the New Standards?" *Special Report: Standards in Electronic Learning,* March 1993, pages 18–19.

1.8: Hawkins, Jan, "Technology and the Organization of Schooling," *Communications of the Association of Computing Machinery,* Volume 36, Number 5, May 1993, pages 30–34.

1.9: Rosenberg, Richard S., *The Social Impact of Computers,* Academic Press, San Diego, California, 1992.

1.10: Hansen, Peter, "New Approaches to Best-Practice Manufacturing: The Role of Transnational Corporations and Implications for Developing Countries," United Nations Center on Transnational Corporations, New York, 1990.

1.11: Hopper, Max D., "Rattling SABRE — New Ways to Compete on Information" in *The Information Infrastructure,* Harvard Business Review Paperback, Number 90078, Harvard Business School Publishing Division, Boston, Massachusetts, 1991, pages 10–17.

1.12: Seymour, Jim, "Changing the Rules of the Game," *PC Magazine,* Volume 13, Number 1, January 11, 1994, pages 97–98.

1.13: Welch, Douglas E., "Companies Need to Invest in Training to Reap Productivity Gains," *InfoWorld,* December 6, 1993, page 60.

1.14 Brynjolfsson, Erik, "The Productivity Paradox of Information Technology" *Communications of the Association for Computing Machinery,* Volume 36, Number 12, December, 1993, pages 66–77.

1.15: Rubin, Richard S., "Save Your Information System From the Experts" in *The Information Infrastructure,* Harvard Business Review Paperback, Number 90078, Harvard Business School Publishing Division, Boston, Massachusetts, 1991, pages 99–101.

1.16: Jaikumar, Ramchandran, "Postindustrial Manufacturing" in *The Information Infrastructure,* Harvard Business Review Paperback, Number 90078, Harvard Business School Publishing Division, Boston, Massachusetts, 1991, pages 53–60.

1.17: Neumann, Peter G., Ed., "Risks to the Public in Computers and Related Systems," *Software Engineering Notes,* Volume 19, Number 2, April 1994, pages 2–13.

1.18: AT&T Quality Steering Committee, "Reengineering Handbook," AT&T Customer Information Center, Indianapolis, Indiana, 1991.

1.19: Hammer, Michael, and Champy, James, *Reengineering the Corporation: A Manifesto for Business Revolution,* HarperCollins Publishers, New York, 1993.

1.20: Hammer, Michael, "Reengineering Work: Don't Automate, Obliterate" in *The Information Infrastructure,* Harvard Business Review Paperback, Number 90078, Harvard Business School Publishing Division, Boston, Massachusetts, 1991, pages 18–26.

Chapter 2
Technocentrism and
Technology Transfer

*It was not fair to spring those nineteenth century technicalities
upon the untutored infant of the sixth and then rail at her because
she couldn't get their drift.*

Mark Twain in *The Connecticut Yankee in King Arthur's
Court,* Chapter XXII, The Holy Fountain.

Chuck and Carla each have a general understanding of the size and nature of their personal technology transfer gauntlets. Anyone caught in a gauntlet usually does. Those outside of Chuck and Carla's gauntlets may not, however, share this understanding. The challenges other people face are often more difficult to understand than your own. This is particularly true of technology transfer challenges because they can vary widely by individual; each person's challenges can be very different from everyone else's challenges.

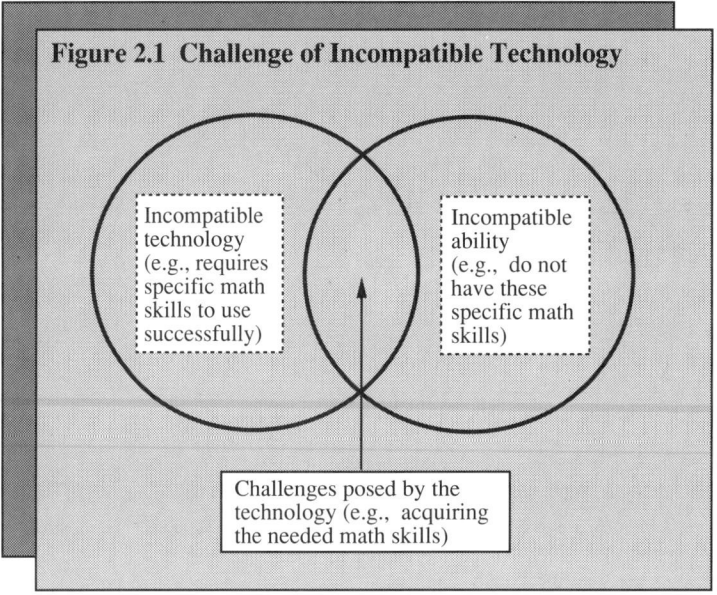

Figure 2.1 Challenge of Incompatible Technology

Incompatible technology (e.g., requires specific math skills to use successfully)

Incompatible ability (e.g., do not have these specific math skills)

Challenges posed by the technology (e.g., acquiring the needed math skills)

Technology transfer challenges vary by individual because they are created through the interaction of the technology and the intended user of the technology. A technology's characteristics interact with a person's life experiences (i.e., a person's beliefs and abilities) to create the challenges posed by that technology. If, for example, use of a particular technology requires mathematical skills that a person does not have, a challenge posed by that technology is acquiring the needed mathematical skills — if the technology is to be successfully transferred (see Figure 2.1). Each person's beliefs and abilities are unique; hence, each interaction with a technology's characteristics is unique. Each person's technology transfer gauntlet is, therefore, unique and created in part by the person who runs it. The gauntlet plus other aspects of this interaction create a person's overall viewpoint or opinion of the technology. The interaction defines a technology's usefulness and ability. Because each interaction or relationship with a technology is unique, each person's viewpoint or opinion of a technology is also unique (there may, however, be some overlap between different viewpoints). A better name for this "viewpoint of a technology" is, in our opinion, *technocentrism* (see Figure 2.2).

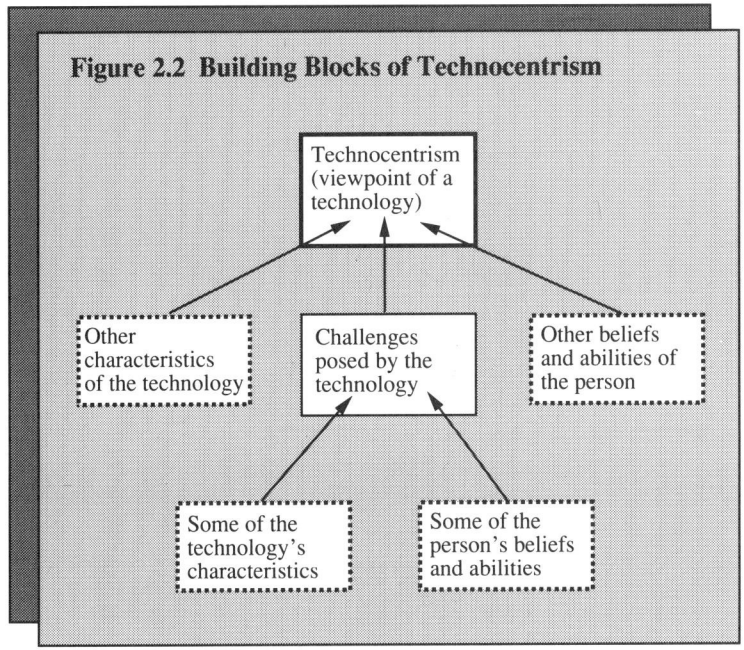

Figure 2.2 Building Blocks of Technocentrism

Why use the term *technocentrism* instead of "viewpoint of a technology"?: because a person's viewpoint of a technology is complex, sometimes emotional, and often (as we discuss at points throughout this book) directly related to the way in which a person defines himself or herself. A person's viewpoint of a technology can relate directly to that person's identity. Consider, for example, the Amish. Can an Amish farmer use computing to maintain farm records and still be "Amish"? Consider the teenager who wants a car. For many teenagers the "car issue" is more central to teenage identity than transportation. In addition, a person often considers his or her viewpoint of a technology to be (just like ethnocentric viewpoints) superior to other viewpoints of that same technology. The word *technocentrism* better conveys these attributes than the phrase "viewpoint of a technology." So what does technocentrism have to do with the technology transfer process? Technocentrism adds tremendous complexity to the problem of technology transfer. If technology transfer is to be successful, the individual challenges in each person's technocentrism must be overcome. A successful technology transfer process must "go through" the technocentrism of each person involved in the project (see Figure 2.3). Recognition of the various technocentrisms at work during a technology transfer project is, therefore, necessary for the project to be successful. The Connecticut Yankee, Hank Morgan, discovered these attributes and the importance of technocentrism during his sixth-century technology transfer project.

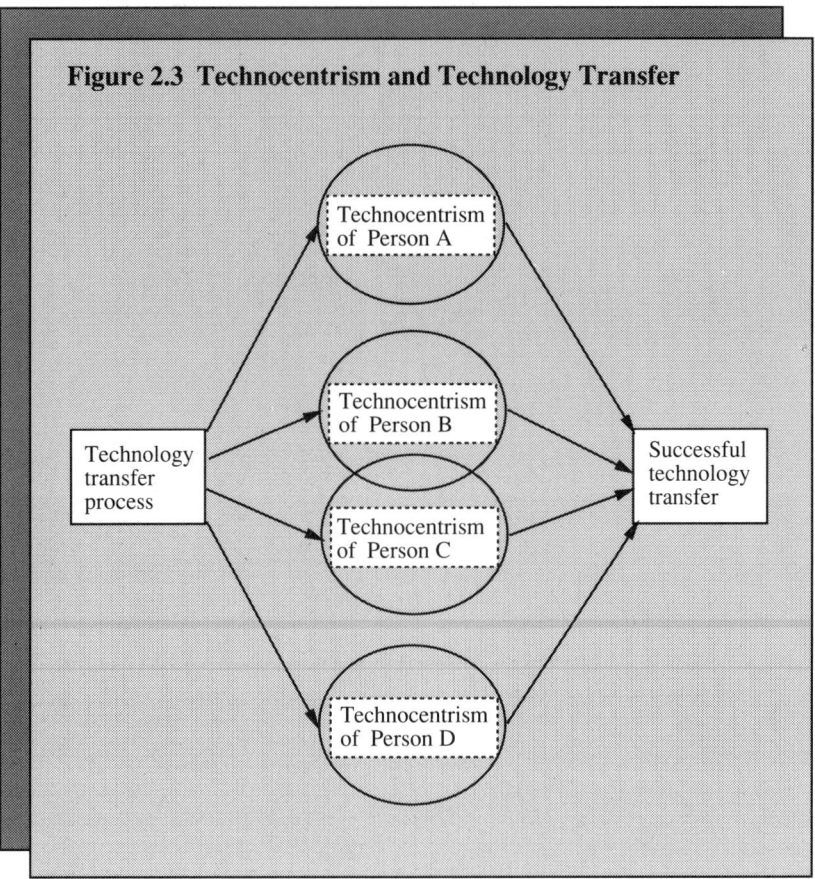

Figure 2.3 Technocentrism and Technology Transfer

Hank Morgan wanted to transfer nineteenth-century technology into the sixth century. Hank's technocentrism judged the nineteenth-century technology logical and useful; the sixth-century members of King Arthur's court, however, judged that same technology to be magical and largely irrelevant. The two technocentrisms shared no common ground. The lack of common ground did not bother the Court much. The Yankee, on the other hand, was absolutely infuriated (to say the least) by the sixth-century view of "his" technology. He wanted sixth-century inhabitants to share his view — his technocentrism — and was deeply annoyed that they did not. His view, after all, was so clearly superior to theirs. He continued to "spring those nineteenth century technicalities" upon sixth-century inhabitants and then "rail at (them) because (they) couldn't get their drift." He only stopped railing against them and their views when he realized, at last, that their views were valid from their "untutored infant" existence. Hank Morgan came to realize that multiple valid views — multiple valid technocentrisms — exist.

This realization does not mean that Hank Morgan even briefly considered adopting the sixth-century view. He still felt that his view was superior; indeed, more than that, his view was right. This realization meant only that Hank agreed that if he were a sixth-century inhabitant, if he had lived his life in sixth-century shoes, he would hold the same view. Nor did this realization make Hank abandon his efforts to bring sixth-century technocentrism into line with his own. The major impact of Hank's realization was a change in strategy to bring about this realignment. Instead of trying to alter their views by force (railing against them, dismissing them and their views as idiotic), Hank would have to alter their views using more surreptitious schemes. He would transfer nineteenth-century technology by starting with the sixth-century view. He would meet sixth-century inhabitants at the dead center of their technocentrism and lead them incrementally to a new view.

The existence of multiple valid technocentrisms may seem a trivial point. It is not. It is the main reason familiarity and proficiency with a technology is developed one person at a time. Each person's beliefs and abilities regarding a particular technology are unique. Anyone interested in the transfer of a particular technology must be prepared to help each intended user overcome the personal challenges the technology presents to each intended user of the technology. The characteristics of each intended user's technocentrism must be understood and accepted as valid. Note also that the acceptance of multiple valid technocentrisms does not mean that all views of a particular technology are correct. Some views are wrong. There is such a thing as fact. If, for example, someone has the view that personal computers are powered by running hamsters, you do not have to agree with it. If you are interested in transferring the technology of workplace computing, all you need do is accept the fact that for this person it is true. You need to remember that for this person, the hamsters are a fact.

Some stories from more modern times may help clarify the concept and nuances of technocentrism. We once saw a demonstration of a piano-playing computer at the Massachusetts Institute of Technology. A musician (who had also programmed the computer) sat at a piano linked to a computer and he played a piece of music. The computer "recorded" this music. Immediately after the programmer-musician finished the piece, he instructed the computer to play it. The computer "played" the piano exactly as the person had (the music came from the piano; the piano keys and pedals moved). The computer-piano combination cost almost $100,000. The programmer-musician explained that this computer-piano would be in great demand by "rich people." They would, he said, invite or pay their favorite pianist to come to their home and record on their computer-piano. The rich people could then enjoy the music at will and forever. At this point, a friend of ours (a businessman specializing in marketing) interrupted: "What is the difference between this computer-piano and a twenty dollar tape recorder?" Our friend went on to say that technically the apparatus was wonderful and that the programmer-musician was a fine musician, but he really could not see anyone — even if they had the money — spending $100,000 for something they could get for twenty dollars. The programmer-musician did not seem to fully understand our friend's point. It seemed so obvious to him that someone would want to spend $100,000. He, consequently, did not provide a convincing answer to our friend's question.

The programmer-musician could not provide an answer because his technocentric view of the computer-piano had little, if any, overlap with our business friend's technocentrism. The two of them were not operating on the same wavelength. Those who work directly with a technology see a technology primarily — and often only — in terms of its inherent capability, its structure. They seek to understand and improve the technology (e.g., to make it go faster, do more things). When this is achieved, the technology is improved, better. For the programmer-musician, the achievement of the computer-piano was enough. It made it worth any price. Our friend's question, however, was centered in a marketing and sales view. His technocentrism only allowed him to see the technology in terms of how he would market and sell it. Which companies and products are the competition? Who might buy this product (see Figure 2.4)?

Our friend had the same basic questions when we recently told him about a computer system that plays back interviews. With this system, a person is first interviewed using a microphone. This interview is "recorded" as a disk file on a computer. People can then access the interview via the Internet. (Internet is essentially a phone line between computers. This phone line allows computers separated by distances to communicate with each other.) The accessed file can then be copied (downloaded) onto a workstation (computer) in your home or office. The interview can then be played back at will and forever using the sound available in the workstation. The interview does not sound like a machine — it sounds like the person who gave the interview. It is a remarkable piece of technology. It is fun to use and the people who developed it are quite proud of it; they should be. Our friend's view of it, however, is that "They have turned $30,000 worth of computer equipment into a ten dollar radio." Which companies and products are the competition? Who might buy this product?

Our friend is not against improvements in technology. Indeed he often wishes that improvements would happen faster (he is very concerned about American competitiveness). What bothers him are improvements without relevance to customers. He often says that many people in many companies are more interested in impressing each other than in impressing customers. Paraphrasing our friend's feelings using "technocentrism," he might say that many people in many companies employ a technocentrism toward their products that leaves the technocentrism of the customer out in the cold. There is no overlap between the potential customer's technocentrism and any technocentrism found in the company developing products for that customer. All our friend wants is technocentrism overlap. Given the number of books and articles on seeing your company and products from your customer's point of view, our friend is not alone in his feeling [2.1–2.3].

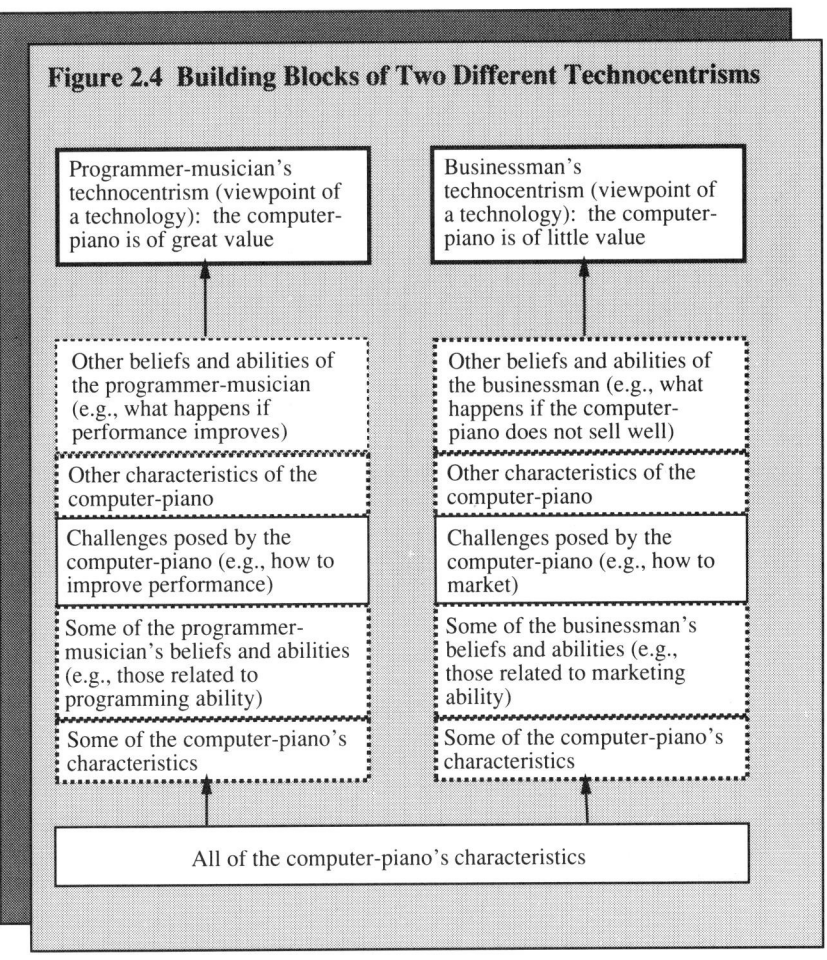

Figure 2.4 Building Blocks of Two Different Technocentrisms

Programmer-musician's technocentrism (viewpoint of a technology): the computer-piano is of great value

Businessman's technocentrism (viewpoint of a technology): the computer-piano is of little value

Other beliefs and abilities of the programmer-musician (e.g., what happens if performance improves)

Other characteristics of the computer-piano

Challenges posed by the computer-piano (e.g., how to improve performance)

Some of the programmer-musician's beliefs and abilities (e.g., those related to programming ability)

Some of the computer-piano's characteristics

Other beliefs and abilities of the businessman (e.g., what happens if the computer-piano does not sell well)

Other characteristics of the computer-piano

Challenges posed by the computer-piano (e.g., how to market)

Some of the businessman's beliefs and abilities (e.g., those related to marketing ability)

Some of the computer-piano's characteristics

All of the computer-piano's characteristics

The friction that our friend feels with those whose technocentrism does not overlap with his (and the friction they feel with him) is not unexpected. That is the nature of technocentrism. One's own view is obviously superior — so obviously superior that everyone should use it. When people do not see the obvious superiority, there is often friction and annoyance. Additional friction and annoyance can result if the nonoverlapping view is also "offensive" (consider the debate over various environmental and health-related policies, technologies, and products). So whose view is right? Whose view is valid? The programmer-musician's? The people who built the "computer-radio?" Our friend's? The customer's? Maybe they all are. Multiple valid technocentrisms exist.

For this reason it is important in any technology transfer process to state and understand the various technocentrisms involved in the project. It is also important to understand clearly which technocentrism is the "decision-criteria" one. Which technocentrism defines the decision rules for the project? The product developer's? The customer's? In this book, it is the customer's technocentrism. In this book, the six technology transfer gauntlets (six decision stages) are created in part and run by the potential business customer of the technology; specifically, the members of Chuck's social service organization and Carla's warehousing company. Note also that customer technocentrism is not monolithic and uniform. All potential customers may not share the same view of any particular technology. This is, as we discuss throughout the book, the situation in both Chuck and Carla's organizations. Throughout the book we discuss the impact of the various technocentrisms on the six stages of the technology transfer process. What are these six stages (stated in terms of workplace computing technology)?

Six Technology Transfer Decision Stages

To transfer successfully workplace computing technology to a business customer, the customer (the intended user of the technology) must decide and choose:

1. Which organizational problem (if any) needs solving.

2. Which organizational resources (if any) are available.

3. Which general solutions (if any) are possible.

4. Which solution technologies (if any) are possible.

5. Which solution products (if any) are possible.

6. Which methods of product use are effective/optimal.

Four attributes of this process (besides noting that business customer technocentrism is operating) require emphasis:

1. The technology-of-interest may not be chosen for use.

2. There are benefits beyond technology transfer.

3. Each decision stage rests soundly on prior stages.

4. The project proceeds and ends without fanfare.

Technology-of-Interest May Not Be Chosen

Sometimes people begin the technology transfer process "knowing" that they are going to use a particular technology-of-interest. Chuck, for example, "knows" that workplace computing will solve his problems. There is nothing wrong with this approach as long as Chuck considers this "knowledge" a hypothesis rather than a fact. Chuck should begin his technology transfer process by saying "I am ready to consider computing as one possible solution to my specific business problems." He should not begin the process with a definite conclusion by saying "I am ready to automate particular organizational tasks." You should use automation if and only if you are certain that use of a computer is the optimal solution to a particular organizational problem. If it is not, then you should not use this particular technology-of-interest. Computers are not silver bullets that stop all organizational monsters. Computers are good solutions to some problems, but they turn into monsters themselves if used incorrectly. Having a technology-of-interest should only mean that you will definitely include it for consideration during the fourth stage of the technology transfer process (Decide and choose which solution technologies, if any, are possible.)

The technology transfer process is, therefore, the process of deciding what technology (if any) is a good solution to your problem. It is the process of systematically thinking about a technology-of-interest and/or other technologies to see if any really do what you think they will. It is the process of trying to build an argument for use of various technologies. Sometimes no such argument can be built. If you cannot build a solid argument, do not use the technology. Successful technology transfer sometimes means not using a technology at all; it means that you have learned that the technology is not for you. (We discuss this in more detail later when addressing the criteria for technology transfer success.)

There Are Benefits Beyond Technology Transfer

The technology transfer process requires that you clearly understand what it is you wish the technology to accomplish. A successful technology transfer process, therefore, begins with a thorough, detailed understanding of current organizational functioning, desired future functioning, and paths to that future functioning. Sound familiar? It should. Much of the technology transfer process (when using business customer technocentrism) is similar to "traditional" processes for successful planned organizational change, development, and improvement (e.g., business process reengineering, strategic planning, total quality management) [2.4–2.7]. As such, organizational improvement usually results from the process even if a new technology is not used. A technology is a potential tool to be used in the ongoing process of upgrading a company's products and functioning.

If this is true, you may ask, why the need to specify another organizational improvement strategy called *technology transfer*? We do this for two reasons:

1. There is often enormous pressure to use a particular technology.

2. A technology must be used responsibly if it is to be effective.

People and businesses often feel enormous pressure to use a particular technology. Workplace computing is a good example. Workplace computing seems to be everywhere and everyone seems to be using it. It is often presented as something that always produces improvements — everywhere, at all times, and independent of who uses it. There are consultants, trade shows, books, magazines, and newspapers devoted to it. Such a "high-pressure" technology may circumvent traditional organizational improvement strategies because it seems so obvious that this technology will improve the company. An improvement process that places equal emphasis on company improvement and evaluation of the technology-of-interest is necessary to keep the technology from being used ineffectively. Effective use of a technology means responsible use. Responsible use means understanding how to use a technology, the impact of using a technology, the side effects of using a technology, when to use it, and when not to use it. Consider the responsible use of antibiotics, chemical pesticides, and nuclear weapons. Traditional organizational improvement methods do not place adequate emphasis on learning to use a technology responsibly. The proposed technology transfer process, on the other hand, incorporates organizational improvement methods.

Each Decision Stage Rests Soundly on Prior Stages

It is also important to note that each technology transfer decision stage rests soundly on prior stages. The major decision made in each stage is, therefore, only as good as the decisions made before it. Consider, for example, how a decision to build a house on sand affects all subsequent construction decisions. If you misidentify the organizational problem, you will probably make incorrect decisions about the general solution and ultimately the products chosen for use. If you misidentify problem solutions, you will probably misidentify ways to implement the solutions (misidentify solution technologies and solution products). The major decision in each stage should, therefore, be a good decision. The decision must incorporate relevant, accurate, complete, and detailed information. It must also be made in a timely way and people must comply with it. If a decision is not timely, key people may lose interest or the "window of opportunity" may be lost. Decisions with which relevant people do not comply are not good decisions. They are not bad decisions either. They are nonexistent decisions.

Project Proceeds and Ends Without Fanfare

So how do you recognize *the* good decision in each technology transfer decision stage? How do you know when technology transfer is complete? How do you know when or if the transfer is a success? There are, unfortunately, no clear-cut signs. There is no one moment of such drama and fanfare in which it is clear to everyone that a good decision has been made, that technology transfer has been a success. The only sign or fanfare is that the argument supporting one particular decision is stronger than the argument supporting others. It is also important to note that the "value" of a decision is often dynamic. It often depends on situational context. Carla, for example, thought the decision to use workplace computing was excellent until she almost went bankrupt. The decision not to put enough lifeboats on the Titanic was a great decision until the ship hit an iceberg. (We discuss good decisions in more detail as we work through the decision stages.) We discuss "successful technology transfer" more thoroughly in Chapter 8. For now, all we will say is that the only reasonable way to measure technology transfer success is from the perspective of expectations. If you use a technology, does its use allow you to do what you expected? Does its use meet the established benchmarks?

It Helps to Have a Good Project Facilitator

Technology transfer is difficult enough when only one person is involved. It is even more difficult when many people (with many different technocentrisms) are involved and their efforts must be coordinated. Technology transfer into the workplace usually involves many people, many technocentrisms, and much coordination. How do you get the project completed and keep it from degenerating into chaos? It helps to have a good project facilitator. It is the facilitator's job to:

- Make sure the project keeps moving.
- Provide feedback on project progress.
- Make sure good decisions are made.
- Make sure relevant people are involved.
- Coordinate the various technocentrisms of relevant people.
- Coordinate, distribute, monitor, and evaluate project resources.
- Coordinate project activities.

To accomplish all of this, the facilitator needs certain skills and qualities. Many of these skills and qualities are the same ones King Arthur wanted in the Knights of the Round Table (that is why there is a knight on the cover of this book). Both knights and facilitators (man or woman) face many challenges and often do battle with all sorts of real and imagined monsters. In the optimal situation, the facilitator should:

- Be familiar with the operation of the company.
- Be familiar with various solution technologies and products.
- Be "comfortable" using information/data.

- Have good interpersonal skills.

- Be committed to seeing the project through to the end.

- Not be easily discouraged.

- Be very organized.

- Be seen as "neutral" by those involved in the project.

- Be able to coordinate the various technocentrisms.

- Be able to develop a sense of shared responsibility/teamwork.

This facilitator does not "lead" the project. The facilitator makes sure that the project gets done. Where do you find such a person? Often within your own company. Sometimes in an outside consulting firm (more about finding consultants in Chapter 9). Occasionally (and with luck) you'll find a facilitator in the sales force of a computer hardware or software vendor. Indeed many of the "old" mainframe salespeople were, in many ways, technology transfer facilitators. "Old" mainframe salespeople are those from the days when large mainframes (computers) were the only computers. Many of these salespeople were famous for knowing as much (and sometimes more) about a company, its business, and its problems than the company's employees. Such salespeople usually made sure that computing was used productively; used where it could help the company and not used where it would have no or negative effect. By and large, however, this "facilitator skill" is no longer routinely found in the workplace computing sales force. It is not that the skill level and interest of the computer salespeople have dropped. Perhaps workplace computing is now too complex — too widely and diversely used — to make such a skill routinely possible. In the workplace computing examples of this book, Chuck will be the facilitator for the social service organization project. Carla will hire a consultant for this job in the warehousing company. Chuck is going to be the facilitator for a number of reasons, not the least of which is lack of money to hire a consultant. Carla will hire a consultant for many reasons, not the least of which is that she does have the money to do so.

Shared Responsibility and Mutual Problem-Solving

Two of the toughest (but critical) aspects of the facilitator's job are coordinating the various technocentrisms of relevant people and developing a sense of shared responsibility/teamwork among the people involved in the project. Shared responsibility is critical to a project such as workplace computing because it involves so many details. No one person can know all the details. No one person can be everywhere at all times making sure that things are getting done. No one person can realistically take sole responsibility for making sure that the chosen computing package is operating effectively. Without shared responsibility, there is no responsibility; things fall through the cracks. Everyone must share the responsibility of making the project work.

All of the relevant people in a workplace computing project have their own view of computing. These various technocentrisms may not be the same. They may not overlap. Carla's Computing Services Department views "effective" computing differently than the warehouse personnel. If the various technocentrisms within a company are not coordinated, then there is the possibility that one department's or one person's view will dominate. A dominant view representing only one department or one person can impair the functioning of other departments or people. It is not unknown, for example, to experience a situation in which computing that improves the administrative functions in a social service organization impairs the work of social workers [2.8]. The automation that makes administrative reporting easier can make keeping case records for social worker purposes worse. The technocentrism of the administrators says computers are great. The social worker's technocentrism says they are a disaster. Both views are correct (that is the nature of technocentrism). When the social workers complain about the negative effects of automation, however, they are often viewed as resistive to automation or noncompliant with the decision. The social worker complaints may be actively invalidated and ignored (e.g., the social workers do not know what they are talking about, the solution is perfect, the automation makes everyone's life better) by those in support of the automation solution chosen by the administration. Everyone usually ends up mad (nonoverlapping technocentrisms often cause friction).

So how do you keep one view (such as the administrator technocentrism) from dominating? How do you develop shared responsibility? The best way to accomplish these tasks is through mutual problem-solving. This is a process (similar to group problem-solving) where those who might be affected by technology transfer solve problems together (we discuss this process in more detail throughout the book). Each person who might use workplace computing, for example, and those who might be affected by its use are part of a problem-solving team. If the warehousing personnel say that a Computing Services Department (CSD) automation decision will negatively affect their functioning (e.g., take them longer to pull orders), then either a compromise is developed, the problem is reconceptualized, or CSD seeks a new solution to its problem (all with the assistance of the facilitator). CSD does not solve its problem by creating problems for other departments.

Chuck does not have a problem using mutual problem-solving. Indeed, for a variety of reasons, everyone in his organization already lets him know their opinions or decisions on everything that happens in the organization. Chuck will use this method largely because he really does not have a choice. Carla is less comfortable with the technique (but will use it). After all, she owns the company. What right do any of "her" employees have to tell her what to do? She knows what is best for the operation of her company. In her heart, she thinks of mutual problem-solving as "mob rule." But mutual problem-solving is no more mob rule than is democracy. What differentiates democracy from mob rule is the existence of standards — the existence of decision constraints — and someone to enforce them. The mutual problem-solving group can do anything it wants to do — within the established constraints. It is the facilitator's job to keep the group functioning within the constraints. The decision constraints are specified during the first decision stage — decide and choose which organizational problem (if any) needs solving.

References

2.1: Whiteley, Richard C., *The Customer-Driven Company: Moving from Talk to Action,* Addison-Wesley Publishing Company, Reading, Massachusetts, 1991.

2.2: Hinton, Tom, *Customer-Focused Quality: What to Do on Monday Morning,* Prentice Hall Publishing Company, Englewood Cliffs, New Jersey, 1994.

2.3: Low, Janet, Malcolm, Bob, and Woolgar, Steve, Ed., "Do Users Get What They Want?" Special Issue of the Association for Computing Machinery's (ACM's) Special Interest Group on Office Automation Systems (SIGOIS) Bulletin, Volume 14, Number 2, December 1993.

2.4: AT&T Quality Steering Committee, "Reengineering Handbook," AT&T Customer Information Center, Indianapolis, Indiana, 1991.

2.5: Hammer, Michael, and Champy, James, *Reengineering the Corporation: A Manifesto for Business Revolution,* HarperCollins Publishers, New York, 1993.

2.6: Stein, Robert E., *The Next Phase of Total Quality Management,* Marcel Dekker, New York, 1994.

2.7: Bank, John, *The Essence of Total Quality Management,* Prentice Hall Publishing Company, Englewood Cliffs, New Jersey, 1992.

2.8: Kagle, Jill Donner, "Record Keeping: Directions for the 1990s," *Social Work,* Volume 38, Number 2, March 1993, pages 190–196.

Chapter 3
Stage One:
Realize the Problem

....realization soon began to supplement knowledge. The mere knowledge of a fact is pale; but when you come to realize your fact, it takes on color.

Mark Twain in *The Connecticut Yankee in King Arthur's Court,* Chapter VI, The Eclipse.

"Chamber, Ludlow Castle, Shropshire, England," photo © copyright Christi Carter, 1993.

The first stage of the technology transfer process is to decide and choose which organizational problem (if any) needs solving. This is not a small task. Reasonable people often come to very different conclusions about what the real problem is and whether anything needs to be done about it at a particular point in time. It is important, however, that the relevant people in the organization come to some agreement on the problem. They must not only collectively "know" what the problem is, but must also collectively "realize" it. Realization must "supplement knowledge." Many a problem has either not been solved or solved incorrectly because people knew what the problem was (could tell you what it was), but did not realize it (did not understand the full implication or drama of the problem). Mark Twain says it is the "difference between hearing of a man being stabbed to the heart, and seeing it done." Realization leads to action (problem-solving) in ways that knowledge does not. Consider the "realization" of many of the world's problems because of television and this realization's impact on policy-making. Seeing war, starvation, or floods on television, for example, is very different from just reading or hearing about them. The path to realizing the organizational problem to be solved (if any) is found by answering the following seven stage one questions.

Stage One Questions:

1. What is the company's mission?

2. What are the company's goals and objectives?

3. What priority is each goal and objective?

4. How effectively is each goal and objective being met?

5. Which ineffectively met goal or objective, if any, needs to be addressed immediately?

6. What are the problems of meeting this goal or objective at each level of the organization?

7. What problem, if any, needs to be addressed and solved immediately?

This first stage of the technology transfer process has surprisingly little to do with any technology. It has more to do with organizational understanding, development, and improvement. The questions in this stage are similar to questions asked in "traditional" organizational improvement processes. In technology transfer, however, answering these questions is a smaller percentage of the entire process. In the technology transfer process, these questions (this first stage) are "getting one's ducks all in a row" questions. In the technology transfer process, this first stage is also the first of a four-stage "speed bump" designed to keep use of any technology-of-interest from careening out of control (see Figure 3.1). The speed bump is necessary to keep the technology from being incorporated before its impact is fully understood.

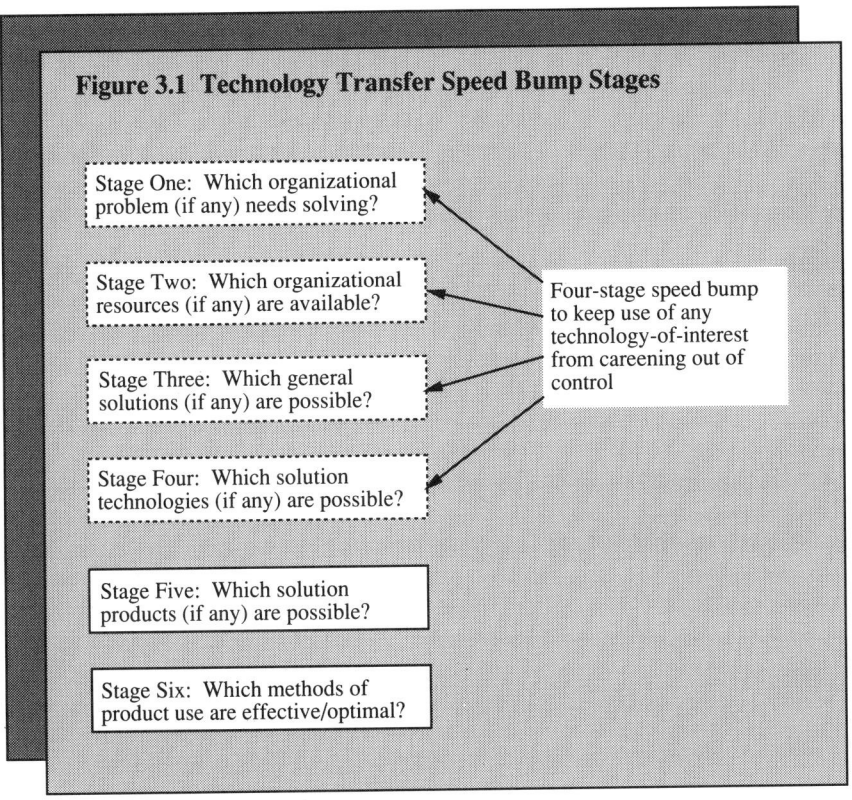

Figure 3.1 Technology Transfer Speed Bump Stages

Stage One: Which organizational problem (if any) needs solving?

Stage Two: Which organizational resources (if any) are available?

Stage Three: Which general solutions (if any) are possible?

Stage Four: Which solution technologies (if any) are possible?

Stage Five: Which solution products (if any) are possible?

Stage Six: Which methods of product use are effective/optimal?

Four-stage speed bump to keep use of any technology-of-interest from careening out of control

Use of a technology-of-interest sometimes careens out of control because of the intense focus placed on it. Often the existence of, access to, or lack of access to a technology is seen as *the* problem [3.1, 3.2]. The obvious solution (without the speed bump) is to alter the existence of, access to, or lack of access to the technology-of-interest. Buy the technology and all the problems are solved. We have all heard people say things such as "The problem is that I do not have a computer and everyone else does," "The Japanese are economically killing us because they are automated and we are not," "My child cannot learn to read without a computer in school and at home," "I cannot write the essay because I do not have a computer at home." Use of workplace computing technology is enormously appealing. Chuck too sees lack of computers in his office as *the* problem. He believes that once he has any kind of computer on everyone's desk, all will be well. He is ready to buy computers — today, now. If he does this, however, without fully understanding the role that computers will play in his organization he could be worse off than he is now. Traditional organizational improvement processes do not adequately address or compensate for the enormous power or appeal of some technologies, the desperate sense of urgency that some people feel about using a particular technology.

It should also be noted that this stage (and indeed the entire technology transfer process) pays enormous attention to the ways in which goals and objectives are met at all levels of the organization (Question 6 — What are the problems of meeting this goal or objective at each level of the organization?). We specify four levels: organizational (system-wide), divisional, departmental, and individual. Different companies may have different names for each level. The choice of goals and objectives at each level forms decision rules and constraints at the next level. The achievement of organizational goals requires a valid relationship between organizational, divisional, departmental, and individual goals and objectives. The more valid, understood, and accepted are goals at all levels and their relationship, the more likely it is that organizational goals will be achieved. Organizational goals and objectives ultimately form decision rules and constraints on individual tasks and behaviors. Conversely, individual tasks and behaviors ultimately determine whether organizational goals and objectives are met. Individual tasks and behaviors ultimately determine the character of organizational functioning. Achievement of individual goals must lead to departmental goal achievement. Departmental goals must lead to divisional goal achievement. Divisional goals must lead to organizational goal achievement. The late Tip O'Neill, former Speaker of the U.S. House of Representatives, said "All politics is local." The same is true of organizational functioning. The organizational interests of individual employees are sometimes local (How does current or future organizational functioning affect me personally?). And organizational functioning is built on the local (individual) interests, efforts, and activities of employees.

Individual goals and objectives are met through the details of an individual's work. Understanding these details is important in any organizational improvement process. It is particularly important in technology transfer because technologies such as workplace computing often affect such details. The detailed ways in which goals and objectives are met by individuals constitute decision rules and constraints for the use of a technology (see Figure 3.2).

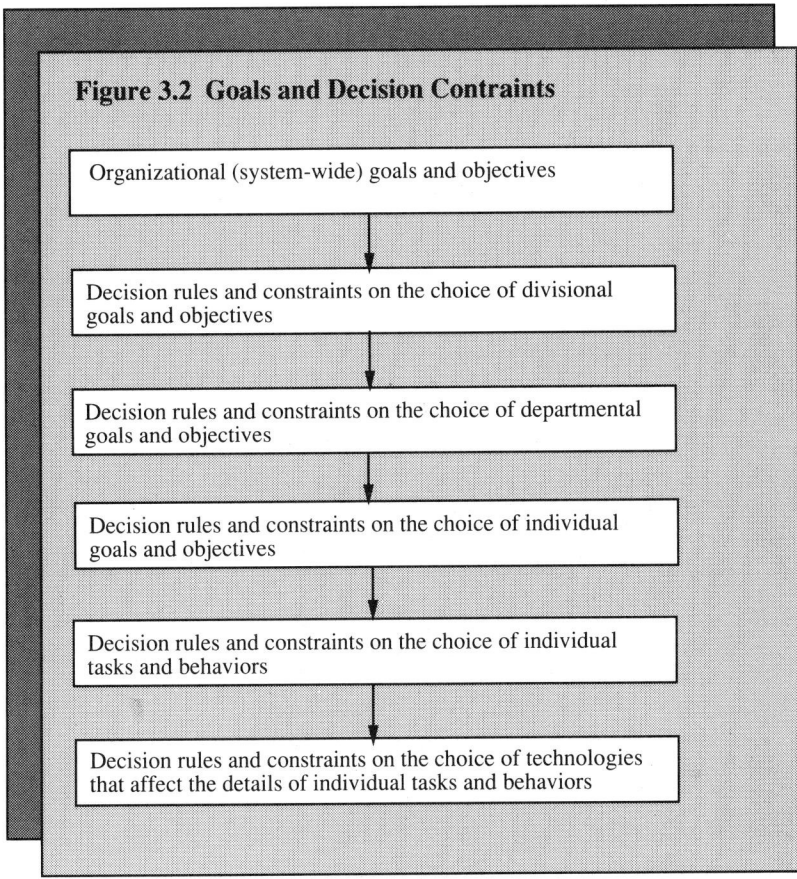

Figure 3.2 Goals and Decision Contraints

It is important to ensure that use of a technology does not impair those detailed individual behaviors and tasks critical to organizational functioning. We have seen some people choose, for example, scheduling or appointment software that decreases the time needed to generate management reports, but increases the time needed for a receptionist to schedule an appointment. Whether this is important or not depends on whether scheduling an appointment rapidly is important to the achievement of organizational goals. Is the organization apt to lose clients or customers because of the increased time involved in scheduling an appointment? The technology transfer facilitator and others involved in the technology transfer process need to know the answer to that question. The problem of one person cannot be solved by creating organizational problems elsewhere; everyone in the organization should be working together as a team.

So what does all this mean for Chuck and Carla? Well, we will move fairly quickly through the first five questions and pick up the story in more detail at Question 6 (problems of meeting goal). We do this primarily because there are many fine books, articles, and seminars on establishing and meeting organizational mission, goals, and objectives. Both companies (for the purposes of this discussion) have good and clear knowledge of their mission, goals, and objectives. The facilitator was quickly able to gain agreement on Questions 1 through 5. The answer to Question 5 (Which ineffectively met goal or objective, if any, needs to be addressed immediately?) is organizational survival. All organizations have a common goal of survival. This is not usually clearly stated in most goal statements, but it is clearly on everyone's mind. Both the social service organization and the warehousing company feel, for different and varied reasons, that going out of business is a real possibility.

Survival as an organization has become a high-priority goal. The relevant people in both organizations could have chosen a more delicate alternative to the word "survive" (e.g., compete more successfully). They chose not to be delicate, preferring to state bluntly what everyone was thinking. Answering Question 6 requires answering four questions:

1. Who are the relevant people for Question 6?

2. What are the stated/presented problems to goal achievement?

3. What are the problems for discussion?

4. What are the identified problems?

The relevant people for Chuck are everyone in the organization. They, for the most part, are going to voice their opinions anyway. So it is best just to include everyone from the beginning. Any interested person from the organization can decide and choose which organizational problem (if any) needs solving. For Carla, the relevant people are everyone in the Computing Services Department (CSD), some of the field engineers associated with the hardware company that sold Carla the machine, assorted other people (e.g., a few people from the warehouse), and Carla. The stated/presented problems to goal achievement are what relevant people say the problems are. The relevant people form the mutual problem-solving group. It is their job to reason together and solve problems together. It is their job to decide what the real (identified) problems are. Some of the stated/presented problems may not, in fact, be close to reality. The facilitator and the relevant people decide on the wording of two questions to determine stated/presented problems. For simplicity, these two organizations have decided — independently and remarkably — to ask the exact same two questions. The facilitator in both Chuck and Carla's organizations asks everyone (the relevant people) to answer the following questions:

1. What is your gut reaction as to why this company is having a tough time surviving? Why, in your opinion, is everything such a struggle?

2.   What in your daily work could be done better? What (with current resources and then if you had more resources) could you do better that would make life less of a struggle for the company as a whole? And for yourself?

The facilitator needs to protect people's anonymity during this information gathering phase. Many people do not want their name associated with this information. For this reason, Chuck will not gather the information himself. He asks two people (a social worker and a secretary) who are respected and trusted by the people in the organization to gather the information. Chuck also develops his own list. All three lists are turned over to a Board of Trustees member. The outside consultant (facilitator) in Carla's organization gathers the information from the relevant people (see Figures 3.3 and 3.4). This information can be gathered in group meetings, individual meetings, or both. It is likely that many people will vent when answering these questions (especially when answering the first question). This is true under any circumstances, but when a company is facing "survival" problems it is especially true. There is nothing particularly wrong with this. What looks like venting may contain valid information (even paranoid people have real enemies) and the venting itself may serve a purpose. What the facilitator should avoid, however, is letting the venting become an endless search for blame.

Finding out how the company got into this situation (who is to blame) is only important if it leads to a way out of the situation. Usually it does not. Figuring out why the Titanic had a large hole in its hull, for example, and who was to blame for it and the lack of lifeboats added no information to the immediate problem of getting people off the sinking ship. In most technology transfer situations it is better just to deal with the situation at hand and leave why (and blame) to the historians and attorneys. The facilitator should also expect a wide range of responses to these two questions — some seemingly insightful and some seemingly stupid. While gathering the information all responses should be treated as valid. Some of the responses that Chuck and Carla got were: Chuck is an idiot, Carla is an idiot, assorted other people are idiots, the social service organization relies too heavily on outside funding, there is no problem, everything is great, assorted people do not care about the company, Carla's machine is a lemon, the software does not work.

The facilitator takes these responses and develops an answer to Question 6.3 (What are the problems for discussion?). These should be problems that have a good chance of being selected as the identified problems. The identified problems are "what is really going on." The facilitator develops this list by going through the list of "presented" problems looking for a common theme and combining this information with additional information the facilitator should have about the company and its environment. Chuck and the board member do this job together. If the facilitator is lucky, one or more of the presented problems is a really good candidate for the identified problem; somebody hit the nail on the head. The facilitator then brings the list back to the relevant people to get agreement on the identified problems (sometimes getting this agreement requires the development of a new list of discussion questions). Question 7 is then answered by the same group of relevant people (What problem, if any, needs to be addressed and solved immediately?). The answer to Question 7 is the problem that the organization tries to solve.

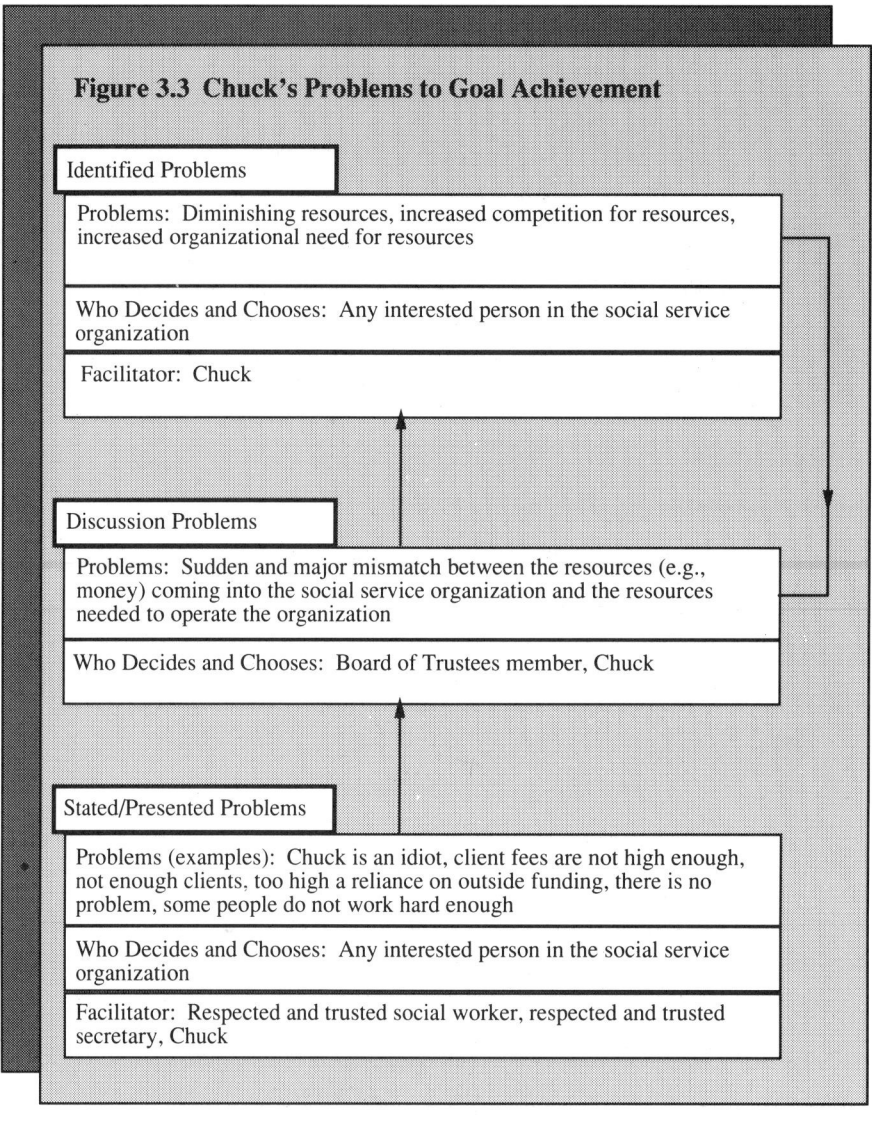

Figure 3.3 Chuck's Problems to Goal Achievement

Identified Problems

Problems: Diminishing resources, increased competition for resources, increased organizational need for resources

Who Decides and Chooses: Any interested person in the social service organization

Facilitator: Chuck

Discussion Problems

Problems: Sudden and major mismatch between the resources (e.g., money) coming into the social service organization and the resources needed to operate the organization

Who Decides and Chooses: Board of Trustees member, Chuck

Stated/Presented Problems

Problems (examples): Chuck is an idiot, client fees are not high enough, not enough clients, too high a reliance on outside funding, there is no problem, some people do not work hard enough

Who Decides and Chooses: Any interested person in the social service organization

Facilitator: Respected and trusted social worker, respected and trusted secretary, Chuck

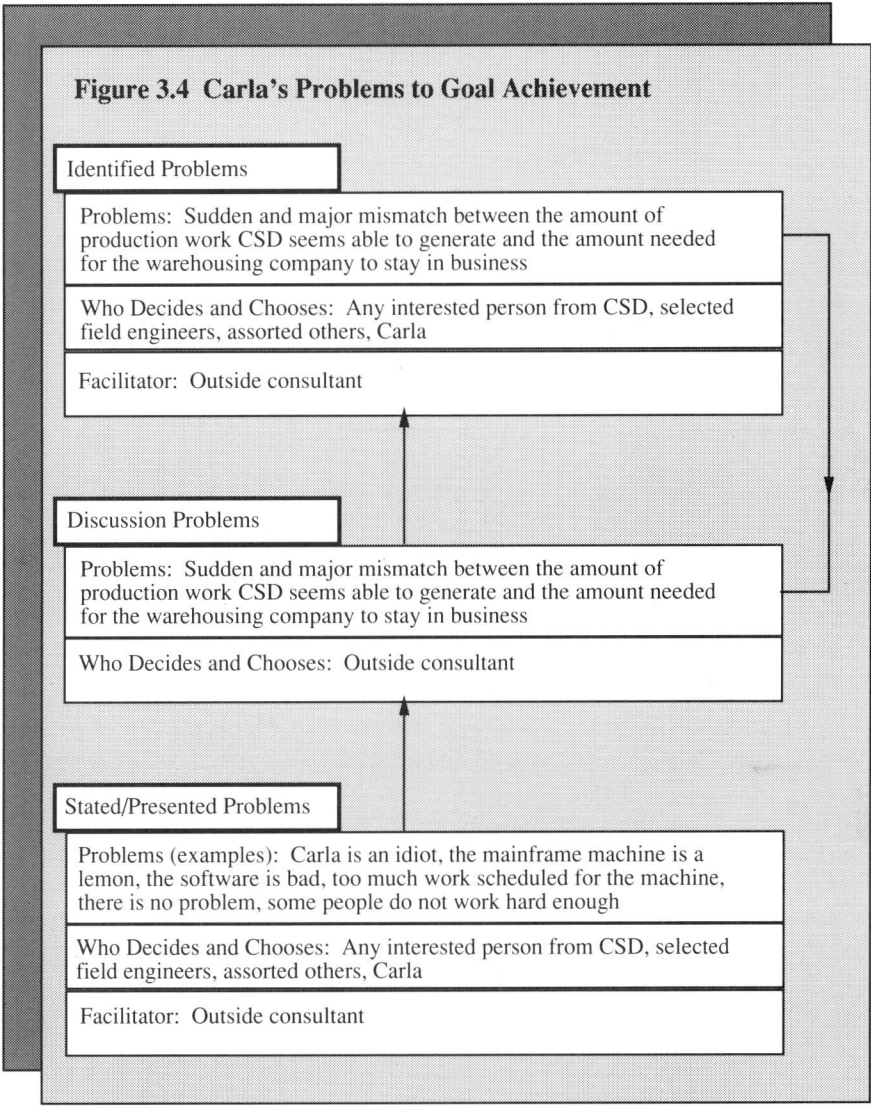

Figure 3.4 Carla's Problems to Goal Achievement

Identified Problems

Problems: Sudden and major mismatch between the amount of production work CSD seems able to generate and the amount needed for the warehousing company to stay in business

Who Decides and Chooses: Any interested person from CSD, selected field engineers, assorted others, Carla

Facilitator: Outside consultant

Discussion Problems

Problems: Sudden and major mismatch between the amount of production work CSD seems able to generate and the amount needed for the warehousing company to stay in business

Who Decides and Chooses: Outside consultant

Stated/Presented Problems

Problems (examples): Carla is an idiot, the mainframe machine is a lemon, the software is bad, too much work scheduled for the machine, there is no problem, some people do not work hard enough

Who Decides and Chooses: Any interested person from CSD, selected field engineers, assorted others, Carla

Facilitator: Outside consultant

The answer to Question 7 for the social service organization is diminishing resources (especially money), increased competition for many of those resources, and increased organizational need for resources. The organization has come to the end of the road for "doing more with less." The situation is not expected to improve in the short term. The answer to Question 7 for the warehousing company is that there is a sudden and major mismatch between the amount of production work the Computing Services

Department seems able to generate and the amount needed for the warehousing company to stay in business. It is important for the identified problem to be stated broadly with no implied solution in its phrasing. Problems are often stated in terms of their solution (as discussed earlier where lack of computing is seen as the problem). This practice limits solution possibilities.

In an ideal world, this stage ends with all relevant people fully realizing the problem facing the company. What sometimes happens, however, is that some people realize the problem, some just have knowledge of the problem, and some people miss the boat completely. Often the best that a facilitator can achieve is a "limited" agreement:

- An agreement that a particular problem is one that the organization will spend time solving; and

- That those people who think it is a bad idea will at least not do anything to disrupt the process.

Sometimes reaching even this limited agreement on an identified problem is difficult and takes every bit of interpersonal skill and expertise a facilitator has. A personality characteristic that is absolutely essential for the facilitator is the ability to know when it is time to "fish or cut bait." We have seen projects planned into oblivion because relevant people could not come to the point where they either did "something" or dropped the process entirely. It is the facilitator's job to keep the process moving; to do what has to be done to get fishing or baiting under way by the people who need to be doing either one or the other. This is no small task.

The relevant people in Chuck's organization have reached a limited agreement. There is a social worker in Chuck's organization named Mary (we will meet her throughout this process) who really thinks spending time on this is silly — at best. She believes — actually she *knows* — that this process is leading toward use of computers and she really thinks they are a waste of time in a social service organization. She agrees, however, not to interfere with the process. Part of the reason that she does agree is because she knows that top management of this organization is behind the process. She knows that the resources available to top management (e.g., firing people, allocating money) will be put behind this endeavor. It is enormously helpful — if not critical — to this technology transfer process (or any organizational improvement process) if top management puts the resources available to it behind the process. Verbal support alone is usually not effective. Most employees are very good at figuring out whether any project is "for real" or just talk. The warehousing company, on the other hand, has come to total realization of the company's immediate problem. Once the problem is identified, it is time to decide what resources, if any, are available for problem solution.

References

3.1: Miller, Jeffrey, "Know When to Fold'Em," *Information Week,* May 5, 1992, "Final Word."

3.2: Gassee, Jean-Louis, "Teachers Before Computers in the Education World," *MacWeek,* Volume 6, Number 33, September 21, 1992, page 42.

Chapter 4
Stage Two:
Thriving Resources

A hermit thriveth best where there be multitudes of pilgrims.

Mark Twain in *The Connecticut Yankee in King Arthur's Court*, Chapter XXI, The Pilgrims.

"Carving on the Cathedral Chartres," photo from the Cooper Union Library Picture Collection.

The second stage of the technology transfer process is to decide and choose which organizational resources (if any) are available. Which resources does the company have now? Which resources can be easily obtained? Pilgrims may be a great source of resources for hermits, but most companies have to look elsewhere (e.g., banks, foundations, venture capitalists, plain old capitalists). Deciding which resources are realistically available before closely examining solutions to a particular problem is essential; it keeps you from seriously considering solutions you cannot afford. It also keeps you from wasting time and should keep you from confusing wishful thinking with serious consideration. It is also fairly common in daily life. When considering housing, for example, people usually make a rough estimate of what they can spend before beginning a detailed examination of specific houses or apartments. When considering college, families usually make a rough estimate of what they can spend before a child sends out college applications or the family visits campuses.

Stage Two Questions:

1. Who are the relevant people for this stage?

2. How much money is available for the project?

3. How much time is available for the project?

4. How much space is available for the project?

5. What information is available for the project?

6. What supplies are available for the project?

7. What structure is available for the project?

8. Which people are available for the project?

What a company can afford to spend — without sacrifice — are its available resources. Such resources are available in the sense that their use does not harm or potentially harm a company's current or future existence. This second stage of the technology transfer process is a critical stage because available and provided project resources shape a company's improvement project (or any other project) just as surely as a house's foundation shapes a house. Why? Because any project rests on the provided resources the way a house rests on its foundation. Company projects and houses resting on inadequate or incorrectly placed (allocated) foundations are unstable. A house's foundation also determines the housing possibilities because once the foundation of a house is in, what the house can be (e.g., how large, the shape) is restricted to a few possibilities. What a company can afford (the amount of available resources) helps to restrict project possibilities in the same way. If, for example, a company can afford $100, it cannot buy a $100,000 computer. Which resources are available definitely shapes the general and specific problem solutions that should be seriously considered. To determine the amount of available resources, eight stage two questions must be answered.

The resources a company has available can be determined with some certainty — if the facilitator is clear on who the relevant people are. The relevant people (those who decide and choose) are those people who control the resources. Because the organizational personnel who control the financial resources, for example, may not be the same people who control supplies, each of the seven remaining questions probably has a different group of relevant people. The relevant people for each question are discussed later (also see Figures 4.1 and 4.2).

Available Money

To assess adequately the amount of available money, three questions must be answered:

1. Is there a cap? (Is there some amount of money above which the company does not want the cost of the project to go?)

2. How much money is available over a short period of time? (How much money is available today from a bank account or a loan?)

3. How much money is available over a long period of time? (What is the cash flow available for the project over time?)

All companies need to control costs — even costs that improve the company. One way to control costs is to place a cap on the amount of money spent on a company improvement project. There are many ways to position this cap. A company can cap the total amount of money spent on a project (e.g., two thousand dollars and not a penny more). A company might also cap the project's daily, weekly, monthly, or annual expenditures. The cap should acknowledge both short-period money (money available over a short period of time) and long-period money (money available over a long period of time). Short-period money is needed to buy the improvement project equipment (e.g., computing equipment) and pay the basic project costs (e.g., cost of using an outside consultant). Long-period money is needed to pay any additional fixed costs the improvement project creates (e.g., cost of maintaining the computer equipment, salaries of any additional personnel). Deciding how much money is available for a company improvement project is not particularly difficult. It is at least no more difficult than deciding how much money is available for any other company expenditure (e.g., salary increases, travel). Companies already have a process in place to answer these questions. The relevant people are easy to identify — those responsible for the financial health of the organization. What these "financial people" *say* the company can spend is what the company *can* spend. There may be people in the organization who think that more or less can be spent, but the decision rests with those assigned responsibility for it.

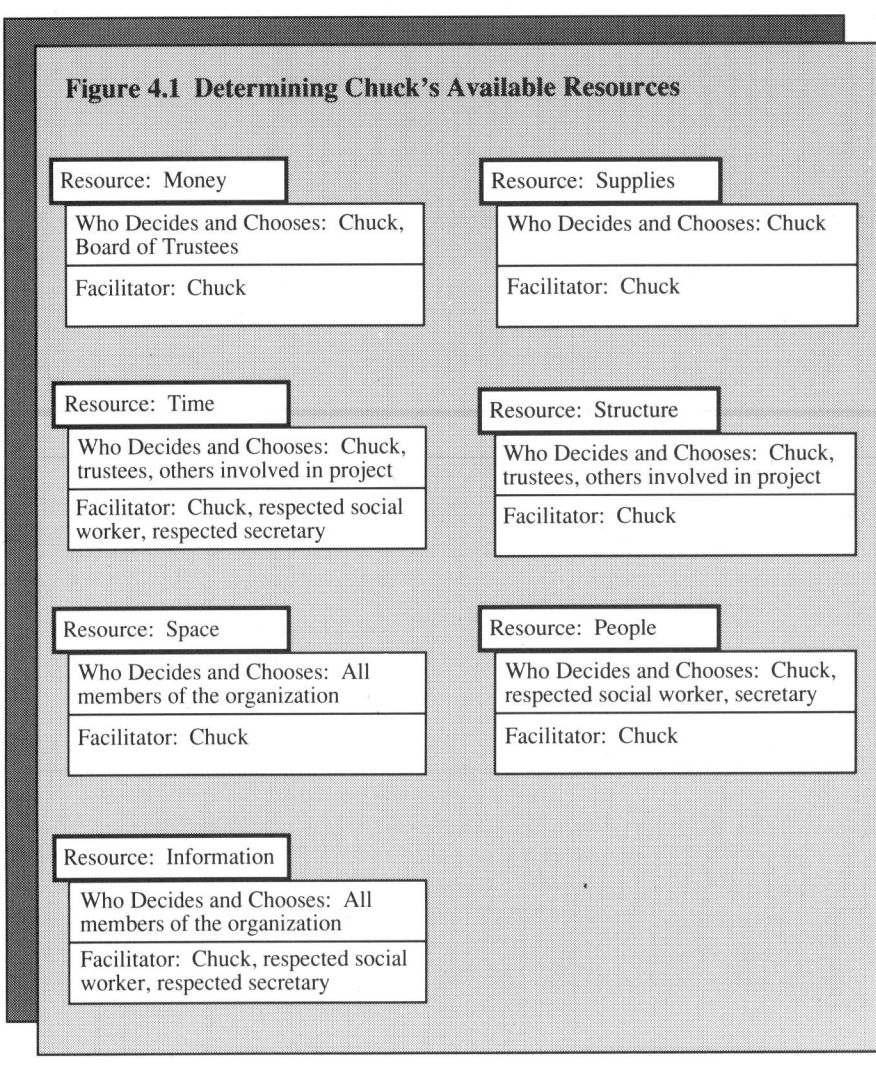

Figure 4.1 Determining Chuck's Available Resources

Resource: Money
Who Decides and Chooses: Chuck, Board of Trustees
Facilitator: Chuck

Resource: Supplies
Who Decides and Chooses: Chuck
Facilitator: Chuck

Resource: Time
Who Decides and Chooses: Chuck, trustees, others involved in project
Facilitator: Chuck, respected social worker, respected secretary

Resource: Structure
Who Decides and Chooses: Chuck, trustees, others involved in project
Facilitator: Chuck

Resource: Space
Who Decides and Chooses: All members of the organization
Facilitator: Chuck

Resource: People
Who Decides and Chooses: Chuck, respected social worker, secretary
Facilitator: Chuck

Resource: Information
Who Decides and Chooses: All members of the organization
Facilitator: Chuck, respected social worker, respected secretary

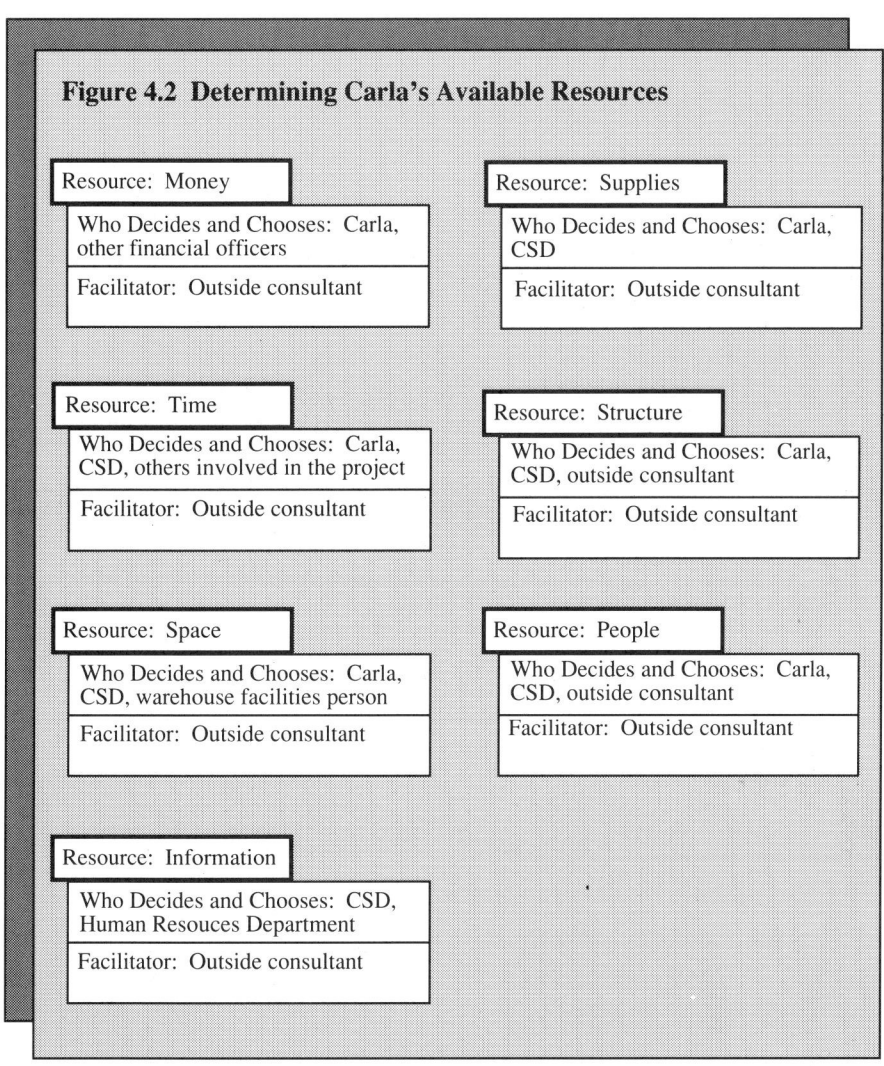

Figure 4.2 Determining Carla's Available Resources

Resource: Money

Who Decides and Chooses: Carla, other financial officers

Facilitator: Outside consultant

Resource: Supplies

Who Decides and Chooses: Carla, CSD

Facilitator: Outside consultant

Resource: Time

Who Decides and Chooses: Carla, CSD, others involved in the project

Facilitator: Outside consultant

Resource: Structure

Who Decides and Chooses: Carla, CSD, outside consultant

Facilitator: Outside consultant

Resource: Space

Who Decides and Chooses: Carla, CSD, warehouse facilities person

Facilitator: Outside consultant

Resource: People

Who Decides and Chooses: Carla, CSD, outside consultant

Facilitator: Outside consultant

Resource: Information

Who Decides and Chooses: CSD, Human Resouces Department

Facilitator: Outside consultant

Chuck and the Board of Trustees are, therefore, the relevant people in his organization. The relevant people for the warehousing company are Carla and the various other financial officers of her company. Chuck has no money from the current operating budget to pay for any improvement project — short or long term. He hopes to get either a grant from a foundation or money from private individuals and figures that $10,000 is the largest amount he can get. Carla's situation is better. Hers is the fortunate company with some cash reserves and a good credit line. She does not, however, really want to spend more than $500,000. She hopes that whatever solution is used to solve her current computer problems does not cost more than she currently spends in the Computing Services Department (CSD) (see Figures 4.3 and 4.4).

Figure 4.3 Chuck's Available Resources

Resource: Money

Cap: Probably $10,000

Short Period: Essentially none - grant dependent
Long Period: Essentially none - grant dependent

Resource: Supplies

Usual assortment of office supplies (e.g., fax machines, copier) at selected times, extra desks, chairs, and bookcases

Resource: Time

Deadline: None - as soon as is reasonably possible
Routine Time: Essentially none - cannot cancel client appointments
Crisis Time: Essentially none - long commutes, second jobs

Resource: Structure

Support from top management, overall organizational enthusiasm for the improvement project, no friction between various organizational divisions and departments

Resource: Space

Working Space: Conference room at selected times, some employee desks and offices
Storage Space: Essentially none

Resource: People

Support from most employees (except Mary), general goodwill between employees, opinion leaders are the respected social worker and secretary, possibility of volunteers to help on project

Resource: Information

Formal Information: Information related to human services and human service organizations
Informal Information: Variety including computer hardware and software

Figure 4.4 Carla's Available Resources

Resource: Money

> Cap: $500,000
>
> Short Period: $500,000
>
> Long Period: Same amount currently spent in CSD

Resource: Time

> Deadline: Six months from today
>
> Routine Time: Essentially none - nothing in CSD is routine anymore
> Crisis Time: Essentially none - already using all crisis time

Resource: Space

> Working Space: Conference room at selected times, two vacant offices, some employee space
> Storage Space: The two vacant offices (if not used as working space), space in warehouses

Resource: Information

> Formal Information: Information related to shipping, warehousing, computer hardware and software
> Informal Information: Variety - at least as far as CSD personnel can remember

Resource: Supplies

> Usual assortment of office supplies (e.g., fax machines, copier), warehousing supplies (e.g., forklifts), and computing supplies (e.g., printer, mainframe) at selected times, extra desks

Resource: Structure

> Support from top management, overall organizational enthusiasm for the improvement project, no friction between various organizational divisions and departments

Resource: People

> Support from all employees, general goodwill between employees, two programmers are the opinion leaders, no real possibility of volunteers to help on the project

Available Time

To determine adequately how much time is available, three questions must be answered because three different kinds of time must be considered. The first question concerns elapsed time, while the second two concern human resource time. These three questions are:

1. Is there a deadline? (Is there some point in time by which the project must be finished?)

2. How much time is available on a routine daily basis?

3. How much time is available on a crisis/emergency basis?

It is important to specify the existence and nature of any deadlines that constrain an improvement project. Does the project have to be completed in six weeks? One year? Why? Is it a firm deadline? If not, how much slippage is possible? By deadline, we mean something outside the general control of the company; something established in whole or in part by people external to the company (e.g., planned sale of the company, mandated compliance with a federal or state law, due date for a specific report). If no such deadline exists, then a company can determine its own time line for the improvement project (such a time line is determined in Stage Three: Which General Solutions, If Any, Are Possible). It is important to determine a time line even if no deadlines exist; the absence of a firm deadline does not mean that an infinite amount of time for the project exists. Without a time line, it is also hard to keep the project a priority. When there are two things to do, one with a defined due date and one without, we almost always work on the one with the defined due date. Many other people seem to do the same thing. The ASAP project may continually be pushed behind projects with a specified due date. In developing the time line, a large project should be divided into smaller self-contained projects that can be completed within short periods of time. A company may have a 5-, 10-, or 20-year plan, but the plan should be divided into smaller plans so that every few months or so there is some measurable improvement and feeling of accomplishment. A frequent feeling of accomplishment is important because it is hard to sustain interest and excitement in a project that seems never to end. The relevant people (the mutual problem-solving group — those who decide and choose) are those who understand any time constraints that the company may be facing. In Chuck's case, the group includes Chuck, the Board of Trustees, and a few others (e.g., the Director of Professional Services). Carla has a fairly well-defined deadline — before her current machine fails completely and forever (six months to a year) and definitely before the maintenance contract on this machine is terminated (in six months). So Carla's deadline is approximately six months from today. Chuck does not have such a well-defined deadline. His deadline is "as soon as is reasonably possible." But Chuck really does not have any more time to waste than Carla.

The second question (how much time is available on a routine daily basis) really asks whether the improvement project can be built into the regular duties of personnel who need to be involved in it. Can it become routine to spend time on this improvement project? Some companies may have already built such time into employee schedules. Some or all employees may already be involved in ongoing quality control meetings, strategic planning meetings, etc. If so, employees may be able to use the "regularly scheduled time" for this improvement project. Companies without such time who can build the current project into employee schedules might want to consider keeping this "regularly scheduled time" permanently. After the current project, there are bound to be other things that need improvement. Another way to ask this second question is "Can the company realistically mandate that each person set aside time each week to work on organizational improvement?" If so, when is this time — during regular working hours or after hours or both? If it is after hours, what incentives are there for an employee to spend this time on company business? If it is after hours, there is often a fine line — or only a few minutes — between routine time and the sort of time addressed in the next question.

This next question (how much time is available on a crisis/emergency basis) asks how long the company can operate as if it were in the midst of a serious crisis or emergency. How long can the company function under enormous stress? How long can the company operate in an extraordinary manner beyond capacity? Working beyond capacity for a short period of time may be possible, but sustained operation at excess capacity endangers product quality, organizational functioning, and company survival (not to mention the mental health and physical health of employees). A company needs to know the answer to this question because sooner or later most improvement projects require some crisis/emergency time (this seems to be especially true when automation is involved). There are two basic forms of crisis/emergency time: during regular working hours and after regular working hours.

In the first case employees suspend most or all regular duties to work on the improvement project. People come to work at the usual time, spend all or most of the day working on the improvement project, and then leave work at the usual time. Little, if any, of the company's usual daily business gets done. The question then becomes "How long can the company go on like that?" On this schedule the employees will not collapse, but the company might. No company can survive long if it does not complete its daily business. Different companies will differ in the length of time they can hold out before collapse. In one company, some or all employees might be able to spend a week or more of regular working hours on the improvement project. In another company the time limit might be two days. In the second form people essentially begin to work two jobs, completing their routine duties during regular working hours and then working long hours on the improvement project after hours. The question then becomes "How long can the employees go on like that?" How long can people continue to work 10-, 15-, or 20-hour days before they wear out? If the people wear out, then the ongoing work of the company stops. Different employees will differ in the length of time they can hold out before collapse. Some employees may be able to work 15- or 20-hour days for no more than one week before they collapse. Other employees, however, may be able to hold out for one month or more.

Who are the relevant people for these two questions? They are the people who might actually work on the improvement project plus those responsible for company production. This latter group should be able to tell the facilitator what production tasks must be completed and when; they should know what the future production schedule looks like. Using this information, the facilitator and the production group need to estimate what time might be available during regular working hours for the improvement project. The facilitator then needs to ask those who might work on the improvement project how much time each can devote to the project during regular working hours. This latter group of relevant people also needs to be asked how much time each might be able to spend on the improvement project after regular working hours and for how long. When individuals who might work on the project report how much time each can spend, they need to be taken at their word. There may be those in the company who think any particular person can spend more time. The facilitator might even think the person can spend more time. But in the end, what really matters is what the person who is being asked to do the work thinks. The facilitator needs to ensure that these time estimates are extremely realistic and conservative. Overestimating can lead to disappointment. In gathering this information, the facilitator must respect the confidentiality of the revealed information. Some people may have reasons for not being able to work longer hours that they may not feel comfortable revealing to the employer (e.g., a second job, an illness).

For these confidentiality reasons, Chuck again asks the two people who gathered information in Stage One (the social worker and the secretary who are respected and trusted by the other employees) to gather time information. Carla's consultant gathers this information in the warehousing company. This facilitator also meets with Carla and various management personnel in her company to determine the production workload. Chuck and some of the Board of Trustees members provide this information for the social service organization. Carla's company does not realistically have any time during regular working hours for the improvement project. Indeed, this company has been working in total crisis mode for months. All relevant employees are already putting in hours of overtime just to keep the company's "body and soul" together. The situation in Chuck's organization is not much better. All of the social workers in the organization are booked solid with clients. To generate "spare time" during regular working hours, the organization would have to cut back on the number of clients seen. This is not realistic for a variety of reasons including budgetary ones. The social service organization does have regularly scheduled quality control and quality improvement meetings that target intervention and treatment strategies used with clients. These meetings cannot be used for the current improvement project, however, because the continual improvement of client intervention and treatment strategies is too important. Finally, it is really not possible to have many people work long hours outside of regular hours for a long period of time. All of the part-time employees have other jobs. Some of the full-time employees have commutes to and from work lasting almost two hours one way. Many full-time employees also hold second part-time jobs.

Available Space

Any improvement project requires space. People need space to work (e.g., room for meetings, office for a consultant) and space to store project material (e.g., computer supplies, books, notes). Determining available work and storage space is fairly straightforward. If the entire organizational space is not very large (e.g., one 3-story building), the facilitator can inventory available space by walking through the building and talking to people. Is there a conference room? Is there space for additional storage? Can people make room in their offices to store material? Can people work on an improvement project at their desks? Some people can and some people cannot. People in a noisy, high-traffic area (e.g., receptionists) may not be able to work on an improvement project at their desks. When people are asked if they have working or storage space in their office or at their desk, they should, once again, be taken at their word. If they say that they have no room, for example, then for all intents and purposes there is no available space — regardless of what anyone else thinks. If the space is large (e.g., two or more buildings, one 20-story building), then the facilitator should also speak to the person in charge of facilities. Speaking to such a person should provide the information faster than having the facilitator roam around the building peering into offices, conference rooms, and closets. Chuck and Carla both have conference rooms that can sometimes be used for the project. Chuck does not, however, see much available storage space or additional work space (every square inch seems to be in use). Some of the social service organization employees say that they can work on the project at their desk and some say they cannot. Carla also has two empty offices in the building that houses the Computing Services Department. There is also some space available at the warehouses, but the warehouses are really not close enough to CSD to make this efficient space to use.

Available Information

Information is also required for any improvement project; *which* information depends on the nature of the project. Every organization is full of information; *which* information depends on the nature of the organization and its employees. Employees acquire some of this information as part of the job. Other information is acquired on one's own time, but may be valuable on an organizational improvement project. Determination of the information (i.e., knowledge, expertise) available within the organization is done in a general sense only. There is no need to administer an array of standardized tests to determine the exact skill range and level of every employee. If the organization is not large, the facilitator can get a general feel for what information is available by chatting with the people who work there. If the organization is large, the facilitator should also speak with the Human Resources Department. The general formal information available in Chuck's social service organization is information related to human services (e.g., intervention and treatment strategies) and human service organizations. To determine the informal information available, Chuck again enlists the aid of the trusted and respected social worker and secretary. Many people develop skills in their personal life (e.g., through hobbies or clubs) that are not routinely used in their jobs. People may also be reluctant for a variety of reasons to tell their boss about these skills. These people may, however, tell their friends at work or a respected and trusted social worker or secretary.

The search for available informal information reveals — much to Chuck's surprise — a variety of somewhat sophisticated skills. It turns out that Chuck's assumption that most of the people in the social service organization are not particularly technical or mathematical is not correct. Some of these discovered skills are not specifically related to workplace computing (e.g., coaching, horticulture, music, real estate, investment). Chuck discovers, however, that one secretary, Susan, is actually very proficient with computers. She gained this proficiency by working with her children on their home computer and through a second part-time job that she has. On most Saturdays, Susan works in a medical office doing billing, letters, and reports using computers. The general formal information available in Carla's organization is information related to shipping, loading and unloading ships, warehousing, distribution, computer software that supports these activities, and computer hardware. There is also a variety of informal information available within CSD — at least as far as the CSD people can remember. The last 18 months have been so grinding that a personal life with hobbies, clubs, etc., is only a distant memory.

Available Supplies

Determining the available supplies (and equipment) is also relatively staightforward. It is (like space and information) determined in the general sense and not the specific. If the organization is not large, the facilitator can inventory available supplies by walking through the building. If the organization is large, then the facilitator should also speak to the people most familiar with the organization's supplies and equipment (e.g., purchasing, stockroom, storage room). Chuck's organization has the usual assortment of supplies found in an office (e.g., fax machine, copier). Nothing can be dedicated entirely to the improvement project, but they can — to some extent — be shared. There are also a few extra desks, chairs, and bookcases stored in the basement. Carla has the same basic assortment of office supplies. She has, in addition, equipment and supplies related to stevedore work and warehousing (e.g., forklifts). She also has an assortment of computing services supplies and equipment (e.g., the current mainframe computer). Like Chuck, none of this can be dedicated to the improvement project, but it can be shared.

Available Structure

Determining the available organizational structure means determining what aspects of the organization are available to support the improvement project. Does the structure exist to support the use of available resources? Does the improvement project, for example, really have the support of top management? As stated in the last chapter, many employees are highly reluctant to actually part with "available" resources unless top management sanctions it. The facilitator also needs to determine if there are any aspects of the relationship between the various organizational divisions and departments that may interfere with the process of an improvement project. Is a certain amount of goodwill available between the various organizational parts or do they act like mortal enemies? Do the secretaries see themselves as adversaries of the social workers? Do the computer programmers refuse to talk to the warehousing personnel? The relevant group for this question includes the facilitator. It is the facilitator's job to make the final decision concerning which organizational structure is available. The facilitator should talk to various members of the organization, but it is the facilitator's job to determine the "reality" of what he or she is told by other members of the relevant group. Are people saying (and possibly believing) that all organizational components really get along when they do not? How serious is top management's stated support of the improvement project? In both Chuck's organization and Carla's organization there is a great deal of organizational structure available for the improvement project. There is genuine support from top management, an overall organizational enthusiasm for an improvement project, and a lack of friction (indeed there is general goodwill) between the various organizational components.

Available People

Determining the available people is similar to determining the available structure. It means, in part, determining the extent to which the improvement project has the support of individual employees. Just because there is general enthusiasm within an organization for the improvement project does not mean that every individual within the organization supports it. Mary (the social worker we mentioned in an earlier chapter) is a good example of this situation. Even though the social service organization as a whole and most of the individuals in the social service organization support an improvement project, Mary does not — not if it means the possible use of computers. Mary will not sabotage the project, but she will probably use every opportunity to object to it. It should be noted, however, that Mary's objection is well intentioned. She really does believe that it is not in the best interest of the clients to spend time and money on workplace automation. Determining why Mary objects to the project is important because the facilitator needs to start thinking about ways to deal with any objections. The facilitator also needs to determine to some extent what the employees who support the improvement project can be counted on to do. Not everyone is going to go out of their way to be helpful. Would Susan, for example (the computer literate secretary in Chuck's office) be willing to teach others about computers? Is there any reason that people would not listen to her? Can people be counted on to learn new skills, if necessary?

Also necessary is a determination of those people in the organization who influence others. Who are the people respected and trusted by others in the organization? Do these "opinion leaders" support the improvement project? In Chuck's case they do — these opinion leaders are the respected and trusted social worker and secretary who have been acting as facilitators for part of this stage. Note that these opinion leaders are often not those who can influence what people do because of the organizational authority they have (e.g., managers, department heads). The facilitator also needs to determine if there are any personal relationships that may interfere with the process of an improvement project. Just because there is a general goodwill between divisions and departments does not mean that everyone in these areas feels the same goodwill. Do some people within the organization act like mortal enemies? Are there any people who would rather be caught dead than agree with each other publicly? Are there any people who will disagree with everyone else just to have something to do — just on general principle? Finally, the facilitator needs to determine if there are any volunteers who might be able to help with the improvement project. Such volunteers might be involved directly in the improvement project or they might complete some employee tasks while other employees work on the improvement project. In Chuck's organization, for example, some members of the Board of Trustees might be able to lend a hand.

The relevant group for this question includes the facilitator (as it did for determining structure). It is the facilitator's job to make the final decision concerning which people are available. How does the facilitator find out all of this information about the people in the organization? If the organization is not large and the facilitator is fairly astute, he or she should have been able to find out this information while determining the availability of the other resources. That is why this question is last. If the facilitator has not been able to find out this information directly, it is likely that the facilitator has been able to identify a few people within the organization who probably do have this information. There are almost always a few people within any organization who are particularly astute at determining the various relationships and politics of the organization. As with determining structure, the facilitator should, therefore, talk to various members of the organization, but it is the facilitator's job to determine the "reality" of what he or she is told by other members of the relevant group. Are people saying (and possibly believing) that everyone in the organization gets along and supports the improvement project when they do not? In Carla's organization there is a great deal of goodwill among most people in the organization. There are the usual tensions found in any organization (especially one operating under extreme stress), but nothing serious. There is also overwhelming support for an improvement project — to the extent that anyone in the organization has any energy left to support anything.

At the end of this stage, the facilitator should have a good idea as to the available resources. The facilitator should also havea good idea as to how the organization really functions (e.g., fairly detailed information on how the work gets done, who really does the work, what the culture is). What if the company does not have many available resources? Well, that situation becomes part of the general problem to which the company seek solutions in the next stage.

Chapter 5
Stage Three:
Schemes and Solutions

It was another of my surreptitious schemes for extinguishing knighthood by making it grotesque and absurd.

Mark Twain in *The Connecticut Yankee in King Arthur's Court,* Chapter XXI, The Pilgrims.

"Gargoyles on Notre Dame Cathedral," Paris, France, photo from the Cooper Union Library Picture Collection.

The third stage of the technology transfer process is to decide and choose general solutions to the problem identified in Stage One. A general solution is a general strategy or a general approach to a problem. Many general solutions are usually possible. Mark Twain's Connecticut Yankee, for example, could have used any number of general approaches (or general schemes) "for extinguishing knighthood." He chose the general approach of "making it grotesque and absurd." But he could also have made it seem immoral or eliminated key knighthood resources (e.g., swords, armor). The Yankee also decided to make the scheme "surreptitious" rather than obvious. One of our favorite computer jokes also illustrates the multiplicity of general solutions or approaches to a problem:

> There are two ways to keep a moving robot from running into people. One way is to program the robot to "see" people and go around them. The other way is to have the robot move slowly in a straight line and have people move out of the robot's way.

OK — so maybe it is not the world's funniest joke, but it does illustrate a point. The identified problem is that robots and people collide. One general solution is to have robots go around people. Another general solution is to have people get out of the robot's way. The chosen general solution dictates the kinds of specific solutions that may be used to solve an identified problem. Different general solutions usually dictate different specific solutions. If, for example, the general solution is to have the robot go around people, then specific solutions involve developing ways for the robot to do this (e.g., ways to make the robot actually see the people in its path). If the alternative general solution is chosen, there is no need to try and develop robotic vision. A better specific solution might be to enroll people in a fitness program so that they have the agility to jump out of the robot's way (as necessary).

The existence of the Connecticut Yankee's scheme (or solution) illustrates an additional point. Throughout the novel, Hank Morgan (the Yankee) functions as a technology transfer facilitator, a facilitator with a technology agenda. Hank wants sixth-century residents to adopt his technologies — his science, his machines, his understanding, and (most importantly) his "politics." Most of all Hank Morgan wants a democracy; he wants knighthood to disappear. Yet he knows that democracy cannot be imposed. It must be chosen. So Hank mentally moves to the center of the sixth-century knighthood view, looks around, and develops a way out. He develops a surreptitious scheme (a strategy) to "lead" people out of their view to his. The scheme is to make knighthood seem grotesque and absurd. All good technology transfer facilitators must be able to do this. They must be able to move to the center of someone else's technocentrism, overlay it with the desired one, determine the similarities and differences, and then develop a way out. They must be able to develop a scheme to lead people from their current view to another one. The scheme usually involves continual slight changes in the person's "original" technocentrism until the facilitator's "desired" technocentrism is achieved (see Figure 5.1).

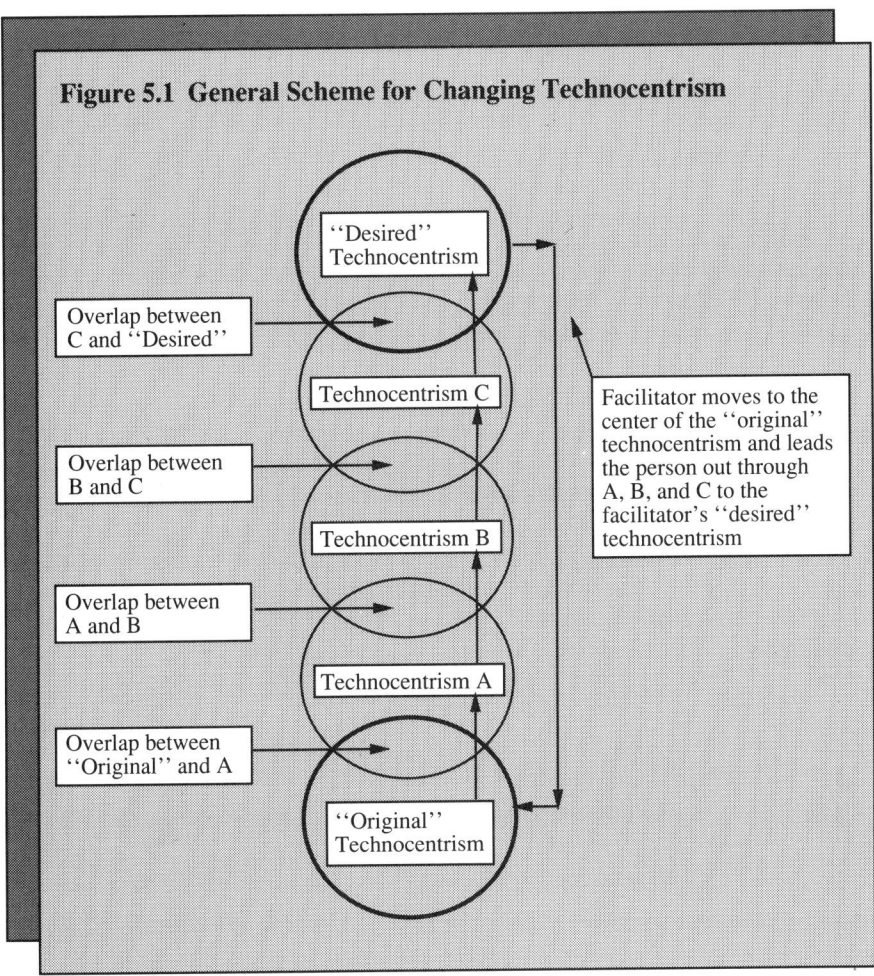

Figure 5.1 General Scheme for Changing Technocentrism

"Desired" Technocentrism

Overlap between C and "Desired"

Technocentrism C

Overlap between B and C

Technocentrism B

Overlap between A and B

Technocentrism A

Overlap between "Original" and A

"Original" Technocentrism

Facilitator moves to the center of the "original" technocentrism and leads the person out through A, B, and C to the facilitator's "desired" technocentrism

Developing a successful scheme is not particularly easy (probably slightly more difficult than doing solid geometry in your head). Sometimes people are aware that they are being led to a new view; the scheme is obvious. Sometimes people are not aware; the scheme is surreptitious. Sometimes the facilitator is well intentioned; the facilitator believes that the new view will improve the person's life. Sometimes the facilitator is not so well intentioned; the new view will mostly benefit the facilitator's life. The good or great facilitators (some with productive agendas and some not) run the gamut from Peace Corps volunteers, agricultural extension agents, teachers/professors and Jesuit missionaries to con artists and drug dealers — with hardware and software sales personnel probably falling somewhere in between. The great technology transfer facilitators are masters at getting people to walk on a "slippery slope." People step on the slope and before they know it, knighthood is dead, they have learned mathematics, they are productively using a computer, they are practicing a new religion, they are growing more corn than they ever have, or they have purchased swamp land in Louisiana.

To determine general solutions to the identified problem, the following 13 stage three questions must be answered.

Stage Three Questions:

1. Who are the relevant people for this stage?

2. What are the stated/presented general solutions to this problem?

3. Did Question 2 provide any information that warrants reevaluation of prior decisions?

4. What are the general solutions for discussion?

5. What are the identified solutions for discussion?

6. Did Question 5 provide any information that warrants reevaluation of prior decisions?

7. What is the estimated cost of each identified solution?

8. Did Question 7 provide any information that warrants reevaluation of prior decisions?

9. Which solutions, if any, are affordable?

10. Did Question 9 provide any information that warrants reevaluation of prior decisions?

(continued on the next panel)

Stage Three Questions (continued):

11. Which solutions, if any, will be chosen for implementation at this time?

12. What is the budget and schedule for the rest of the project?

13. Did Question 12 provide any information that warrants reevaluation of prior decisions?

There is much that is similar between the implementation of this stage and the implementation of Stage One (decide and choose the problem). A key difference, however, is that this stage clearly looks both forward and backward at the same time. It looks backward through the use of feedback loops: Questions 3, 6, 8, 10, and 13. These questions ask whether anything was found that might make a company want to reevaluate something that was decided earlier. Sometimes the search for general solutions turns up new information or creates a slight reconceptualization of the identified problem. In such cases, it is wise to reevaluate prior decisions. This stage looks forward through the development of a ballpark estimate of the cost involved with implementing identified solutions. Developing such an estimate invariably involves guessing (hypothesizing) which technologies might be used in each identified solution. Any technology-of-interest will probably be considered. It is important to remember that at this point any technology-of-interest is only one of the technologies to be considered for use, not *the* technology for use.

Using both the ballpark estimate and the list of available resources developed in Stage Two, a company then determines which solutions are affordable. What should a company do if it cannot afford the solution it wants? The same thing any company or anyone does when it cannot afford any other project or product. A company can, for example, attempt to scale down the solution — choose a smaller project (e.g., automate only one department instead of every department). A company can also start a more rigorous savings plan or it can forget the whole thing. A company must not, however, proceed with a solution it cannot afford. Why? Because life with any project (e.g., automation, home improvement) you cannot afford is a nightmare. Such a project either absorbs all your resources leaving none for other activities or ends when only partly finished.

So how do Chuck and Carla answer the questions in this stage? The relevant people are those who were involved in Stage One. The Stage One facilitator strategy is also the same (the same outside consultant for Carla; the same social worker and secretary in Chuck's organization). The search for presented solutions turns up an assortment of possibilities: fire Chuck, fire Carla, fire assorted other people, open up completely new sources of funding in Chuck's organization (e.g., start a profit-making company that feeds the nonprofit), merge with other organizations, shut down, decrease staff, increase Chuck's ability to compete for existing resources, increase capacity on Carla's computer, decrease the machine's workload, and outsource all the work done by the Computing Services Department (CSD). This last presented solution is possible because CSD operates like a small, stand-alone company within Carla's larger operation. It provides products and services to other components of the company. Many companies are really constructed of smaller, stand-alone units. The facilitators also contact similar organizations to see if they have experienced similar problems and, if so, how they solved them. A list of discussion solutions is developed the same way a list of discussion problems was developed in Stage One. After discussion with relevant people, the following identified solutions emerge for Chuck: to increase competitiveness for the diminishing outside resources and share some resources with other similar organizations (e.g., form a consortium, bulk buying of items). For Carla, the identified solution is to increase the CSD's ability to produce the needed work (workload cannot be reduced) (see Figures 5.2 and 5.3).

Next the company develops a ballpark estimate of any identified solution's cost. Doing this requires that seven resource questions be answered. These are the same ones that were answered in Stage Two. The questions are:

1. How much money is required to complete the project successfully?

2. How much time is required to complete the project successfully (e.g., 100 person-hours, 3000 person-hours)?

3. How much and what type of space is required to complete the project successfully (e.g., 200 square feet of storage space, raised floor, and climate-controlled room)?

4. What information is required to complete the project successfully (e.g., current functioning information, difference between computers)?

5. How much and what type of supplies are required to complete the project successfully (e.g., 200 pens, 150 pads of paper, 2 computers, network cables, modems, 8 file cabinets, 2 tables for work)?

6. What structure is required to complete the project successfully (e.g., unrestricted flow of information between those involved in the project, self-discipline, motivation)?

7. How many people, which people, and what personality characteristics are required to complete the project successfully (e.g., all members of the accounting department, temporary employees, self-disciplined people)?

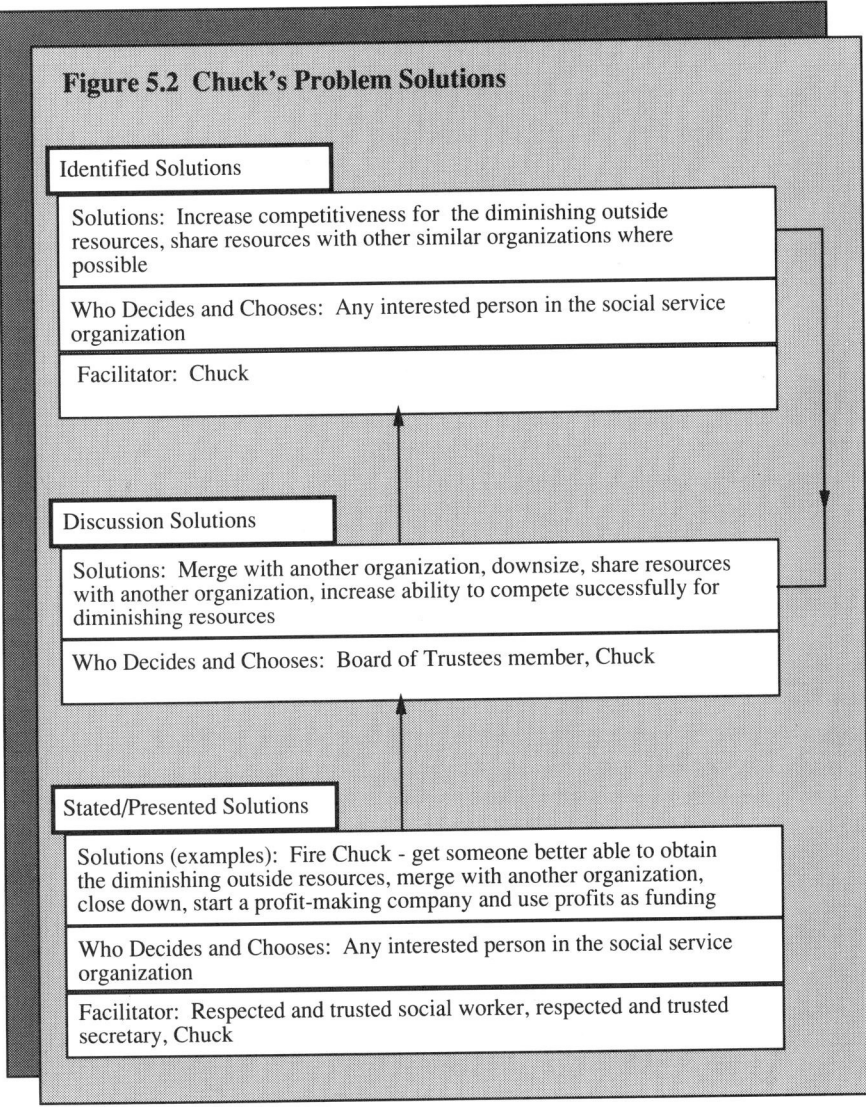

Figure 5.2 Chuck's Problem Solutions

Identified Solutions

Solutions: Increase competitiveness for the diminishing outside resources, share resources with other similar organizations where possible

Who Decides and Chooses: Any interested person in the social service organization

Facilitator: Chuck

Discussion Solutions

Solutions: Merge with another organization, downsize, share resources with another organization, increase ability to compete successfully for diminishing resources

Who Decides and Chooses: Board of Trustees member, Chuck

Stated/Presented Solutions

Solutions (examples): Fire Chuck - get someone better able to obtain the diminishing outside resources, merge with another organization, close down, start a profit-making company and use profits as funding

Who Decides and Chooses: Any interested person in the social service organization

Facilitator: Respected and trusted social worker, respected and trusted secretary, Chuck

When developing the ballpark estimate, both obvious costs (e.g., purchase price of a computer, purchase price of a car) and hidden costs (e.g., computer maintenance costs, cost of annual car insurance) must be calculated and included.

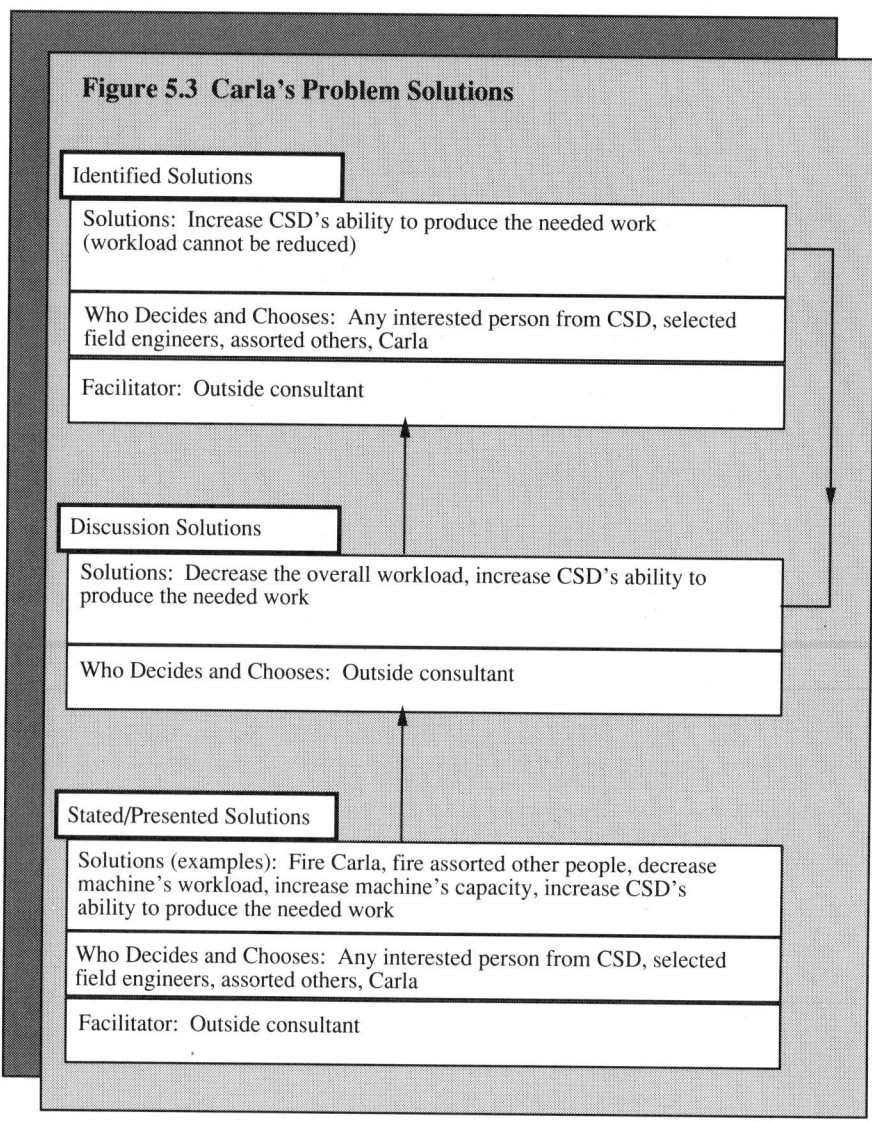

Figure 5.3 Carla's Problem Solutions

Identified Solutions

Solutions: Increase CSD's ability to produce the needed work (workload cannot be reduced)

Who Decides and Chooses: Any interested person from CSD, selected field engineers, assorted others, Carla

Facilitator: Outside consultant

Discussion Solutions

Solutions: Decrease the overall workload, increase CSD's ability to produce the needed work

Who Decides and Chooses: Outside consultant

Stated/Presented Solutions

Solutions (examples): Fire Carla, fire assorted other people, decrease machine's workload, increase machine's capacity, increase CSD's ability to produce the needed work

Who Decides and Chooses: Any interested person from CSD, selected field engineers, assorted others, Carla

Facilitator: Outside consultant

Hidden Costs

What might be the hidden costs when automation is chosen as part of a company improvement project? A common hidden cost of workplace automation that is often overlooked, in our opinion, is the cost of preparing the company's information and information processing for automation. The analogy to this preparation is preparing the surface of a broken cup before using glue to mend it. Unless the broken surface of the cup is prepared (e.g., cleaned, smoothed) the glue will not be effective. The surface of the broken cup must be able to use or work with the glue for the glue to be effective. Unless a company is prepared, automation may not be effective. Company information and processes must be able to use or work with automation for automation to be effective. Just as glue is most effective with a certain type of surface, automation is most effective with certain types of information and information processing procedures.

What do preparation hidden costs look like? Well, Chuck has some good examples of information preparation costs. One is the automation of vacation accrual and absence information. At first glance the automation of absence reporting and recording seems straightforward. Sometimes it is and sometimes it is not. It is not the case for Chuck. For Chuck, the effective automation of such information will require time-consuming reevaluation and greater specification of written personnel policies. Such reevaluation and increased specification will probably require policy decisions and approval of new policies. Why? Because the personnel policies in Chuck's agency state that any employee with more than 20 years of service has one year of sick time. It is not, however, clearly stated in the personnel policies what amount of time is defined by "one year" (e.g., 365 days or the number of work days in one calendar year?). Policies also do not define the calculation of "years of service" since a "break-in-service" is not defined. That is, policies do not state whether a person who worked for the agency for five years, quit, and then returned to the agency one year later had five years of service at the time of rehire or none.

In Chuck's agency, vacation and sick day accrual and use have been handled primarily on a case-by-case basis. Each employee came to some agreement with the organization about years of service. Before automated recording of accrual and absence information can take place, however, these terms must be quantified. The entry of automated absence information is not time-consuming, but the quantification of these and other terms can be. Clearly the policy problems could have been solved in the absence of automation, but there was no motivation to do so (everyone seemed happy with this situation). Automation often illuminates inconsistences, ambiguities, exceptions, and fuzziness. Automation demands a certain amount of precision and, therefore, imprecision must be all but eliminated before automating. Automation also requires general rules since it is not as cost effective to maintain multiple, unique, case-by-case, exception information. One can imagine the case where everyone in an organization accrues sick and vacation time according to a different rule. Automation can help keep track of the unique information, but much of the capability and utility of automation would be lost. In many ways, this preparation is really reducing organizational problems and processes to their "simplest" form. Attempted automation of nonreduced problems and processes can result in technical problems that are not easily solved.

Money and Time Costs

What are some of the other common costs of any technology transfer or improvement project? Well, money is fairly obvious. A company needs money to purchase products. Time is also required for the successful completion of any technology transfer project because people often need time to "go to school" and "do their homework." Every transfer project has a large educational and decision-making component. On an automation project, time is necessary for personnel to consider and evaluate: current organizational processes, competing problem solution technologies, competing automation products, competing automation vendors, and ways to incorporate the chosen automation products into daily life. People need time to learn to use new hardware and software. People also need time to "translate" current workplace procedures into new procedures using chosen computer products. A company also needs to calculate and incorporate the future time and money cost of training new employees in the use of chosen computer products. A company with high turnover may find that they have higher training costs in an automated environment than in a nonautomated one. Each new hire may need fairly extensive training to use the automated system productively. Another potential source of increased costs are the product upgrades (improvements) developed by hardware and software manufacturers. A company may decide to purchase these upgrades to remain current with the technology. It is not always necessary to do this, however, any more than it is necessary to purchase a new car or fax machine every time a new model is released. Use of the upgrade does not necessarily increase productivity. The upgrade may not be "improved" enough to offset the time, effort, and money needed to find, install, and become familiar with the upgrade. We also know people who seem never to do any productive work because they spend so much time working with the upgrades (waiting for them, reading about them, purchasing them, installing them, learning to use them). They seem to spend all of their time getting ready to work and no time actually doing any meaningful work. ("Once I install this upgrade, I can really settle down and get some work done.") Sometimes, however, the new car, fax machine, or computer upgrade is a wise investment. A company should, therefore, estimate and incorporate the cost of purchasing and using some upgrades. Chuck also needs to estimate the amount of time and money that will be needed to locate and secure outside funding for the improvement project. Carla is beginning to think that she needs to allow money and time for someone to write customized payroll and warehousing software.

Space Cost

All automation projects require physical space. These space requirements are of two types: storage space and working space. Storage space is used to store project materials during and after project completion. Automation project storage space includes storage for manuals, computer parts, computer-generated information, printer paper, vendor information, project notes, and many other items. Working space is the area where project work is done and purchased products are used in daily life. Automation project working space includes but is not limited to a place to hold meetings, a place to learn to use the computer, and a place to use the computer after the automation project is completed. Sometimes automation project work space is daily work space with no "fence" to prevent interruptions. We have, for example, known more than one secretary who tried to learn word processing either alone (with a manual or tutorial), with colleagues, or with a paid instructor and during the training sessions had to answer telephones, take messages, and answer questions. These poor secretaries were forever trying to figure out where they were before the interruption. You cannot learn calculus, algebra, or French with constant interruptions and you cannot learn about computers with interruptions either. It should be noted that a lack of space resources can drain time resources since time is wasted moving equipment and trying to figure out where you were before an interruption.

Automated companies sometimes require more storage and working space than nonautomated companies. Employees in automated companies may for additional desks, tables, need this space and bookcases. The employees may need additional supplies and equipment for a desktop computer (sometimes an employee's current desk has enough stuff on it), to store the pounds of paper produced by a computer, to check the information on the pounds of paper for errors and inconsistencies, to store the manuals that usually accompany computer products, and to store disks, tapes, or diskettes. Carla is beginning to think that she will need to construct a bigger or additional machine room to house her computer or computers.

Remaining Costs

Having adequate information of the right kind is also important to the success of any improvement project. In an automation project such required information includes but is not limited to differences between computer models and products, differences between local computer vendors, and the current state of organizational functioning. In addition to good information, those people involved in the project also need good information processing and decision-making skills. The most important structure cost/requirement is keeping the solution implementation a high priority. All too often solution implementation gets lost in the daily activity of trying to keep "the plates spinning."

In thinking about the ballpark costs of the identified solutions, many things become clear to Chuck, Carla, and the relevant people in each organization. Some of this clarity makes them rethink prior decisions. Carla and the others have more than enough money, but they have no time. The current machine may not last long enough for them to put a solution in place. Yet as they rethink the situation, they feel there is no other option but to proceed as quickly as possible. Chuck and the others have come to realize that the total automation project they envisioned at the beginning is financially just not possible. They discussed whether what they could afford would really make a difference. Chuck and others in the organization more seriously considered merging with another organization or closing down. They then decided that even a scaled down project would sufficiently increase their competitiveness. They are beginning to think of ways to divide the total automation project into smaller, self-contained, self-sufficient projects that will lead to the total project. It is better to have one small project (e.g., word processing on one personal computer) working completely and self-sufficiently than one large project only half working. This is especially true if you run on a very tight budget like Chuck. If you cannot afford to automate all the tasks you want to automate at one time (and few companies can), then divide the desired automation into smaller projects that you complete one at a time.

This division of a desired project into smaller ones should be familiar to any homeowner. Few people can afford to make all desired home renovations at one time. What most homeowners do is devise a plan that divides the desired renovations into smaller projects that can be completed as time and money allow. Few homeowners, for example, redo the kitchen, replace the roof, and install new windows concurrently. Most will complete one project at a time. If the homeowners try to complete all the renovations at one time, they might run out of money (or sanity or both) before any one project is finished. If you divide the automation project into smaller projects, make sure that each smaller project (project component) is done well and is guided by an overall master plan. Also make sure that existing components are working well, are stable, and are fully used before adding another. Adding components too quickly may result in a half-used system. People using automation need time to stabilize and adapt to the changes brought about by automation. If you keep adding components before existing ones have stabilized and are fully used, then employees are always working under conditions of transition and crisis. This wears people out.

Project Notebook

Detailing the budget and schedule for the rest of the project is best done by starting an official project notebook. The purpose of the notebook is to remind project participants of the tasks to be done, why a certain course of action was taken (sometimes it is hard to remember after some time has passed), problems encountered (so they can be anticipated the next time), and problem solutions (so project participants do not have to reinvent solutions each time a problem recurs). The "life span" of the notebook is usually the "life span" of the project. Although different companies may include different pieces of information in their notebook, we recommend that at least the following information be included:

1. Original budget

2. Budget revisions

3. Current working budget

4. Original schedule

5. Schedule revisions

6. Current working schedule

7. List of "Things to Do" to meet schedule

8. List of people working on the project

9. List of people responsible for project tasks and activities

10. Information used in decision-making (e.g., product information)

11. List of key decisions and reasoning for each decision

12. List of problems encountered and solutions

13. General project progress notes (can include minutes from meetings)

14. Inventory of purchased products

A few things need to be pointed out at this time. One is that technology transfer is like an iceberg. Much of it is not "visible" — it is done before the technology arrives. We wish also to emphasize that all decisions in this technology transfer process are made under some uncertainty. The facilitator can only move the process along if he or she is one who does not need "complete information" or absolute certainty to act. Another point or issue is how much time it takes to complete the six stages. Possibly not as much as you think. The process can move fairly quickly. Some of the questions can, in fact, be answered in a matter of minutes or hours. People can take days to answer them if they want, but the questions can be answered quickly (a company should try and stick with quickly). We have all been in three hour-meetings where absolutely nothing was accomplished. Some of us have also been lucky enough to be in half-hour meetings where dozens of meaningful decisions were made.

What is certain, however, is that the more complex and massive the project, the greater the cost. Small, straightforward workplace automation projects should not cost much and can often be inexpensively completed in a few weeks. Large workplace automation projects can have enormous costs and require millions of dollars, many people, and many years. It takes time to do things right. Sometimes, however, people manage to spend more time to do it wrong. There does appear, however, to be an optimum amount of resources to spend on any workplace improvement project. If inadequate resources are committed, then a good automation project is not possible. If too many resources are committed, then the project is wasteful and not cost effective. All of this sounds very neat and tidy, but in real life (something we all try to avoid) it is not. Each technology transfer stage may go through many iterations. The process is full of twists and turns, dead ends, false leads, occasional despair, and lots of coffee.

Chapter 6
Stage Four:
Technologies -
Insolent and Grand

The priests opposed both my fire and life insurance, on the ground that it was an insolent attempt to hinder the decrees of God;

Mark Twain in *The Connecticut Yankee in King Arthur's Court*, Chapter XXX, The Tragedy of the Manor House.

"Moreton Corbet Castle, Shropshire, England (destroyed by fire)," photo © copyright Christi Carter, 1993.

The fourth stage of the technology transfer process is to decide and choose technologies that can be used to implement the identified solution. This stage is essentially the same as Stage Three — except that technologies rather than general solutions are the focus. The decision about which technology to use is often difficult. Some technologies present enormous unintended challenges to people. The unintended challenges of technologies can be more troublesome and difficult than those posed by some paradigms (as discussed in Chapter 1). Persistent, incompatible paradigms challenge the way you think. Persistent, incompatible technologies challenge both the way you think and the way you live. Technologies "force" us to behaviors in ways that paradigms do not. Technologies force us to fully understand and live by our paradigms — by what we believe. Consider the high-tech life support systems available in hospitals today. Decisions regarding use or nonuse of such technologies force us, both as individuals and as a society, to clearly understand and live by our paradigms concerning the nature of human life. Such technologies demand behaviors in ways that discussions of conflicting paradigms do not.

Some people, therefore, have philosophical objections to the use of some technologies. Sometimes these objections are based on a person's religious beliefs. The priests opposed the Connecticut Yankee's "fire and life insurance" on religious philosophical grounds — "on the ground that it was an insolent attempt to hinder the decrees of God." The Amish and Mennonites also oppose use of many technologies on religious grounds. Religious grounds preclude the use of some medical technologies by some Christian Scientists. This philosophical objection is not usually to the technology *per se,* but to what the technology represents or makes possible. The technocentrism of the Amish, Mennonites, and Christian Scientists with respect to some technologies incorporates either a harmful or unacceptable effect when the technology is used. Sometimes people or groups whose technocentrism precludes use of a technology are reasonably content if others do not hold the same view. Sometimes a group wishes its technocentrism to be codified as *the* technocentrism as did the sixth-century priests in Mark Twain's novel.

The harmful or unacceptable effects found in some people's technocentrisms are not just related to religious beliefs. Philosophical objections to a technology can be economically based. A person may believe, for example, that use of a technology will eliminate his or her job and that acquired job skills will become irrelevant to society. The person will be forcibly unemployed with little chance of gaining similar employment. Such a person philosophically objects to the technology. However, as with religious objections, economically based objections are not usually to the technology *per se,* but to what the technology represents or makes possible. The technocentrism incorporates the possibility (or the certainty) that use of the technology will cause severe personal and family poverty (maybe even starvation). The nineteenth century Luddites are the classic example of this technocentrism [6.1]. They are also the classic example of the tendency to invalidate this particular technocentrism completely. The Luddite croppers and weavers objected to automated textile mills because such mills made them and their skills irrelevant to the textile industry. Being irrelevant to the textile industry meant you and your family starved. Yet the Luddites themselves and the term came to mean someone who almost "stupidly" objects to technology. Someone who objects to progress and is an unreasonable obstructionist. Life is not that simple. The Luddite technocentrism of yesterday was valid; many of the Luddites did starve to one degree or another. But the technocentrism of the mill owners who wanted automated textile mills was equally valid. Mill owners viewed that same textile technology as necessary for company survival. There is no doubt that automated textile manufacturing helped the textile industry and probably society as a whole. There is no doubt that once one mill started using automation, it changed the textile industry as a whole. As soon as one mill began to use automation, the standards for textile productivity changed. To keep from going out of business altogether, textile manufacturers without automation had to change to automation. And if most of the textile manufacturers went out of business, everyone involved (not just the croppers and weavers) would starve.

So what does this mean for choosing technologies? It means, for one, that the "decision-criteria" technocentrism must be stated. There will probably be many valid technocentrisms at work in any technology transfer project. The decision-criteria technocentrism (the one that drives the project) must, therefore, be clearly stated (see Figure 6.1). This decision-criteria technocentrism is almost always the organization's. The technology chosen for use will be the one that helps the organization as a whole regardless of what it does for any one division, department, or individual. The facilitator should try to develop some overlap between the decision-criteria technocentrism and others, but if this cannot be accomplished then the decision-criteria technocentrism (alone and without overlap) forms the reference point for the project. The existence of many valid technocentrisms also means that the facilitator will have to bear primary responsibility for developing the list of possible solution technologies. Relevant people will be asked for input, but the facilitator must keep in mind that some people may have a vested personal interest in the choice of one technology over another. The facilitator needs to ensure that the stated decision-criteria technocentrism remains the decision-criteria one. Last, but by no means least, we hope that in choosing technologies, a company does not invalidate the "Luddite technocentrism." People do lose jobs to technology. It is bad enough to lose your job; it is worse to have your feelings about the situation invalidated. It is worse to be considered unreasonable or an obstructionist for objecting to a technology that will probably result in personal unemployment. Most people who lose their job to technology do not find another "equal" job regardless of how many jobs the technology itself creates. Life is not usually "just fine" afterward. Some people pay an enormous price for societal change. Transitions to a better or worse societal condition always chew up somebody. This is true for any and all change. Some pay this price voluntarily. Most do not. But even those who pay voluntarily should not also be expected to be gleeful about it.

To determine the technologies that can be used to implement the identified solution, 15 stage four questions must be answered.

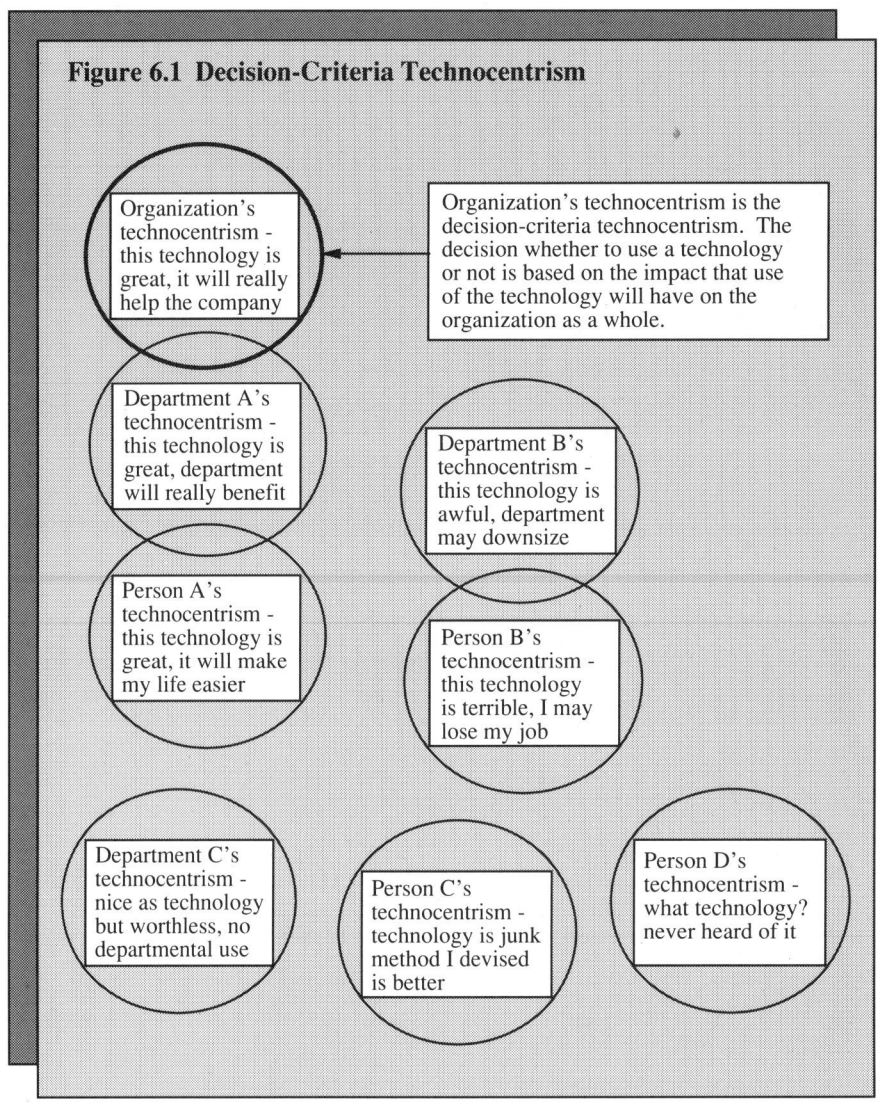

Figure 6.1 Decision-Criteria Technocentrism

Organization's technocentrism - this technology is great, it will really help the company

Organization's technocentrism is the decision-criteria technocentrism. The decision whether to use a technology or not is based on the impact that use of the technology will have on the organization as a whole.

Department A's technocentrism - this technology is great, department will really benefit

Department B's technocentrism - this technology is awful, department may downsize

Person A's technocentrism - this technology is great, it will make my life easier

Person B's technocentrism - this technology is terrible, I may lose my job

Department C's technocentrism - nice as technology but worthless, no departmental use

Person C's technocentrism - technology is junk method I devised is better

Person D's technocentrism - what technology? never heard of it

The relevant people for this stage are the same ones identified in Stage One. Information on the specific tasks that need to be accomplished to implement the identified solution is gathered using the same basic strategy employed in the other stages. Identifying the specific tasks is necessary to develop a list of possible solution technologies. Every technology has certain characteristics that make it good for

Stage Four Questions:

1. Who are the relevant people for this stage?

2. What are the specific tasks (problems to be solved) of the identified solution?

3. Did Question 2 provide any information that warrants reevaluation of prior decisions?

4. What technologies might be used to complete these specific tasks (solve these specific identified solution problems)?

5. Did Question 4 provide any information that warrants reevaluation of prior decisions?

6. What are the technologies for discussion?

7. What are the identified technologies?

8. Did Question 7 provide any information that warrants reevaluation of prior decisions?

9. What is the cost of each identified technology?

10. Did Question 9 provide any information that warrants reevaluation of prior decisions?

11. Which technologies, if any, are affordable?

12. Did Question 11 provide any information that warrants reevaluation of prior decisions?

13. Which technology, if any, will be chosen for use at this time?

14. What is the budget and schedule for the rest of the project?

15. Did Question 14 provide any information that warrants reevaluation of prior decisions?

solving certain types of problems and bad for solving others. Computers, for example, are good solutions if the problem concerning a particular organizational function is actually one of speed, precision, or the inability to store, relate, and manipulate large amounts of information (there is more information about computer characteristics in Chapters 10 and 11). The development of the specific tasks requires a close look at the "solution situation" to see in detail what is really needed to occur to allow the identified solution to solve the identified problem.

After the specific tasks involved in the identified solution are listed, it is time to list possible solution technologies. As stated earlier, this is primarily the facilitator's job. Relevant people should, however, be asked their "gut reaction." Considered technologies should include existing (old) technologies the company currently uses, old technologies that the company does not use (but that others do), and new technologies. The technology-of-interest should be considered, but it should not enter the process as the favorite. Automation, therefore, is only one of several technologies that could be investigated by an organization as a possible solution to a particular problem. The consideration of old technologies helps a company avoid the purchase of new technologies that it does not need. This consideration is another technology transfer speed bump designed to keep people from blindly rushing into the purchase of a new technology. A company's problem may be more efficiently and effectively solved without the use of a new technology. During this stage, the specific effect of each technology's characteristics on problem solutions is stated. When a company chooses to use a technology, it does not mean that the technology is perfect (hardly anything in life ever is). All it means is that the technology solves the organizational problem (meets the stated criteria) at least as well as any other considered technology. It should also be noted that the problem solution may require integrated use of many technologies. Any technology (including automation technology) is merely a tool to help people complete tasks and achieve objectives. Use of the wrong tool does not solve a problem and can lead to additional problems. If no existing technologies seem promising, then a company's primary option is to make do with what it has until a useful technology is developed (that could be a long wait). Under no circumstances should a company use a technology that does not produce benefits for that company — regardless of how many other industries are using that particular technology.

So what do Chuck and Carla realize during this stage? Well, Chuck and other members of the social service organization have come to a more detailed understanding of why this organization is having problems competing for existing resources. What seems to be happening is that requests for information (e.g., from funding sources) have become more important and this organization is less able to meet the information requests than many other organizations. Chuck has come to realize that many people (including many of his funding sources) view the rapid production of precise information about the organization and client services as a sign of overall competence. It is no longer enough just to provide good client service; an organization must be able to prove in a variety of ways to a variety of people that good client service is a reality. Demonstrating competence and knowledge about the social service organization has become as much a part of the organization's product as successful family therapy. Unless Chuck demonstrates competence to his funding sources, he will not get funded. Unless he meets the funding sources deadlines and responds rapidly to any question they choose to ask, he will not get funding. Without such outside funding this organization may go out of business.

Why has the organization suddenly become less able to meet information requests than many other organizations? A variety of possibilities exist. Maybe it is Chuck's data and information skill level. All the data in the world will not help Chuck demonstrate organizational competence if he does not know how to use organizational data. Automated data analysis will not help anyone who does not know how to organize and interpret data. Maybe organizational competence cannot be demonstrated because the organization is not competent. Automated reporting will not help a company that has nothing positive to report. Or maybe Chuck is producing poorly written or incomplete reports. Could automation help in this case? Is some form of automation a potential solution technology to the problem of poorly written papers and, if so, what form of automation would solve the problem? The answer depends on why the reports are poorly written or incomplete. What is the real nature of the problem?

If documents are poor because Chuck knows little about either the art of writing or the subject matter, then automation such as word processing will have little effect on the problem. If the problem is that Chuck does not have skills to do the job, automation is not the answer. Additional training is, however, a possible answer. If Chuck is producing a poor written product because time constraints prevent revision of work, then automation such as word processing (using a computer to create written documents) is a possible answer. In such a case the purchase of a computer with a word processing application is a potential direct solution to the problem, but it is not the only potential solution to a "time-constraint" problem. Noncomputer solutions exist. One noncomputer solution is to decrease Chuck's workload. A workload decrease should allow Chuck to spend more time revising documents. Another noncomputer solution is to increase Chuck's secretarial support. All the secretaries in the social service organization are currently assigned to at least three people. Maybe one secretary should be devoted entirely to support the tasks that Chuck needs to complete. With such secretarial support, it should be possible to have a document typed as soon as Chuck is finished writing or revising it. This shorter turnaround time should allow more revisions to be completed.

How would use of a word processing package solve a time-constraint problem? Should automation be chosen as the solution technology instead of the other possibilities? Answering this question requires, in part, an understanding of the characteristics of word processing. How, specifically, will these characteristics affect the preparation of reports? What, if any, are the advantages of using word processing over the use of a typewriter or pad-and-paper to produce documents? One advantage is easy and time-saving document formatting. By using word processing it is possible to change the appearance of a paper (e.g., size of margins, type of print) without retyping the entire paper. Text (e.g., sentences, paragraphs, whole pages) can also be moved around a document without retyping. The ability to move text easily in a document using a word processing package allows a writer to reorganize the presentation of ideas in a paper more easily than can be done using the more traditional "cut-and-paste" method. This capability can result in a higher quality, more organized, first draft of a document than when a nonautomated technique is used (e.g., typewriter, pen and paper). The amount of written work produced can also increase when using word processing because a document is essentially complete as soon as it is composed. A printer can print the computer-stored document as soon as on-line composition is complete. There is no lag time for typing. With word processing, a writer can revise, and revise, and revise relatively quickly and easily making it more likely that a writer *will* revise, and revise, and revise. Writing with word processing can, therefore, help to eliminate the barrier of "not having enough time to write a quality report." Writing with word processing can also help eliminate other barriers as well. Many word processing packages can also create a table of contents, check for spelling mistakes, and check for grammatical errors. We have known people so unsure of their spelling ability or grammar that they hesitated to write. When they began writing using a word processing package with a spelling checker, however, they gained enormous self-confidence. They no longer fear severe embarrassment from misspelled words. As a side effect of using the spelling checker, their spelling has actually improved also.

Carla has come to realize what many members of CSD and the field engineers knew for months: The mainframe computer hardware and software are operating way beyond capacity for sustained periods of time. Computer hardware — like any machine — has a top performance level. Carla's mainframe has been operating at top performance level almost constantly. This is the primary reason for the almost constant machine failures. A car, for example, may be able to go 100 miles an hour — occasionally. If you operate the car at this top speed continually, it will quickly wear out. Carla should never have purchased the particular mainframe machine she now has. It is the wrong machine for the work that CSD needs to produce. The purchase of the machine seemed to be a good deal at the time; it was not. The software is also being asked to operate beyond capacity. Most of the software is (as we mentioned in an earlier chapter) a kludged-together mess. This mess makes the software error-prone and slow. Until now, Carla had considered this software kludging process to be a cost-effective method of adding software capability; it was not. New, improved, smoothly functioning software is needed in all areas (e.g., payroll, invoices).

So what do Chuck and Carla conclude about their respective problem and solution situations? In Chuck's organization, it is decided that the basic problem is a "time-constraint" one. There is simply too little time. There is not enough time to write reports. There is not enough to time to sort and organize the data needed to produce the reports. The organization chooses automation as a solution technology. At this point, Chuck still thinks that he will automate everything: budget, case records, memos, reports, scheduling, etc. The organization also decides to try another method for reducing organizational costs: forming a consortium with other similar organizations to share resources (e.g., bulk buying). Diminishing outside funding is not as problematic if the organization can reduce expenditures (see Figure 6.2). In Carla's organization, they also decide to use workplace automation as a solution technology. Only this time they will buy a machine that meets CSD's needs. They are also beginning to think that they might purchase two machines (each to serve as the other's backup). Maybe one machine can be located in the current CSD building and be used primarily for payroll. The other machine might be located in a new facility in a warehouse much closer to the docks (closer to the source and destination of most warehousing information than is CSD). This machine might be used primarily to handle warehousing information. Totally new software packages need to be purchased or written (see Figure 6.3).

Once the solution technologies are identified, an organization needs to decide which, if any, are affordable by asking the same seven cost questions asked in earlier stages. This analysis makes Chuck realize that the large-scale project he wants is not affordable at this point. He cannot automate everything. He may only be able to automate report preparation using word processing and spreadsheet software. The automation of data needed to prepare the reports (e.g., case records) may have to wait. During the affordability analysis both Chuck and Carla evaluate the "people" cost of using automation. They find themselves wondering if automation will make some people's jobs obsolete. Each estimates, however, that no job will become obsolete. There will be no layoffs as a result of automation. But what does happen in each case (and usually happens in others) is that there is a point where Chuck and Carla take a good hard look at company personnel and wonder if the desired change in organizational functioning can occur with them. Do current employees have the skill set necessary to "pull off" the changes required in the improvement project? Or are they "in the way"? You certainly do not need introduction of a new technology to have such questions asked. Even on good organizational days, we have each been in meetings where the expressions on our colleagues faces (and maybe our own) looked uncomfortably similar to the expressions of those in the movie *Lifeboat* or the survivors of the Donner party.

When you get this look on your face, however, you should keep a few things in mind. One is, if you are thinking this about others, others are probably thinking this about you. The second is that people with better skill sets may not exist. Whatever you think of the people in your organization, they may be the best there is. An endless search for something better may be a waste of time. Finally, maybe current organizational personnel are completely capable of handling the project. A workplace computing project is really not so different from other projects anyway. The truth is that the automation process is like any other activity that requires both common sense and planning. Most people already have all the information processing (thinking) and

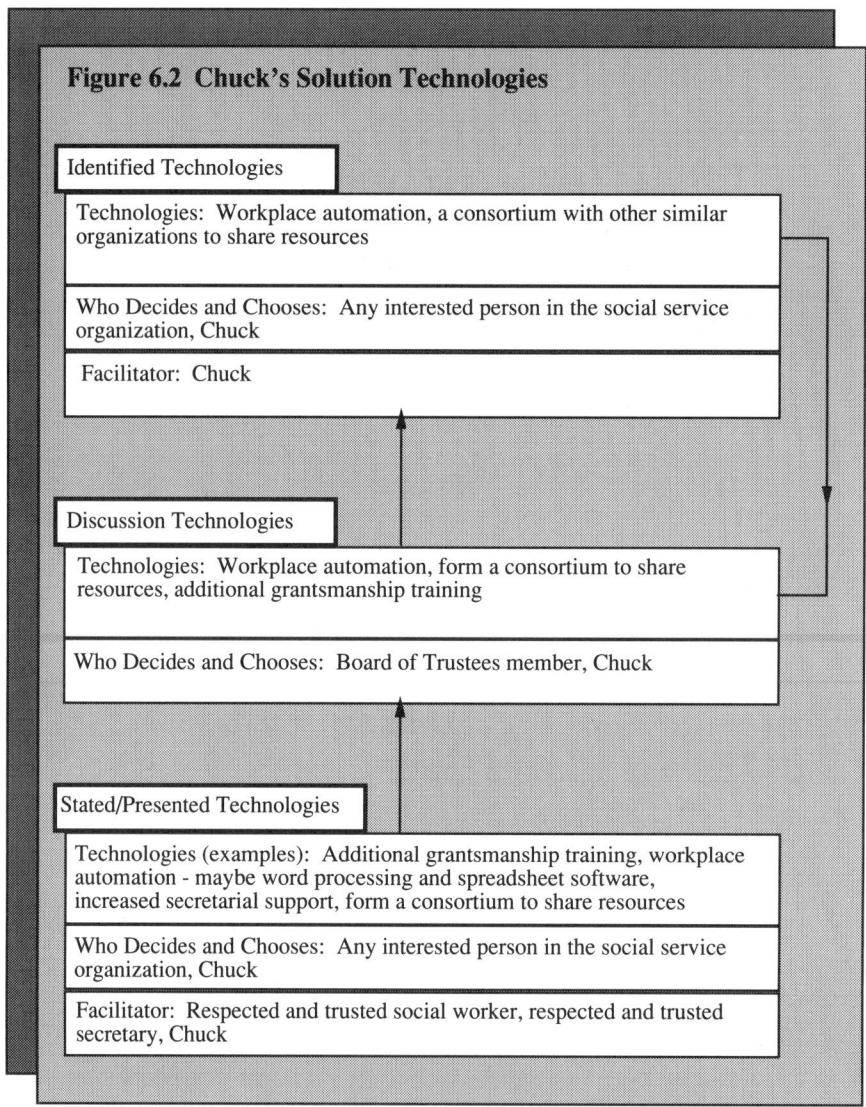

Figure 6.2 Chuck's Solution Technologies

Identified Technologies

Technologies: Workplace automation, a consortium with other similar organizations to share resources

Who Decides and Chooses: Any interested person in the social service organization, Chuck

Facilitator: Chuck

Discussion Technologies

Technologies: Workplace automation, form a consortium to share resources, additional grantsmanship training

Who Decides and Chooses: Board of Trustees member, Chuck

Stated/Presented Technologies

Technologies (examples): Additional grantsmanship training, workplace automation - maybe word processing and spreadsheet software, increased secretarial support, form a consortium to share resources

Who Decides and Chooses: Any interested person in the social service organization, Chuck

Facilitator: Respected and trusted social worker, respected and trusted secretary, Chuck

decision strategies necessary to choose and use automation. Some people are surprised to hear this. These same individuals seem to believe that when they enter the world of computers they must learn a new way of thinking. They seem to believe that when dealing with computers all previous knowledge and skills are of little use and relevance. If you have the information processing and decision skills to buy a car, you have the ones needed to complete an automation project successfully. If you know how to do your job well (if you thoroughly understand your job), you have the

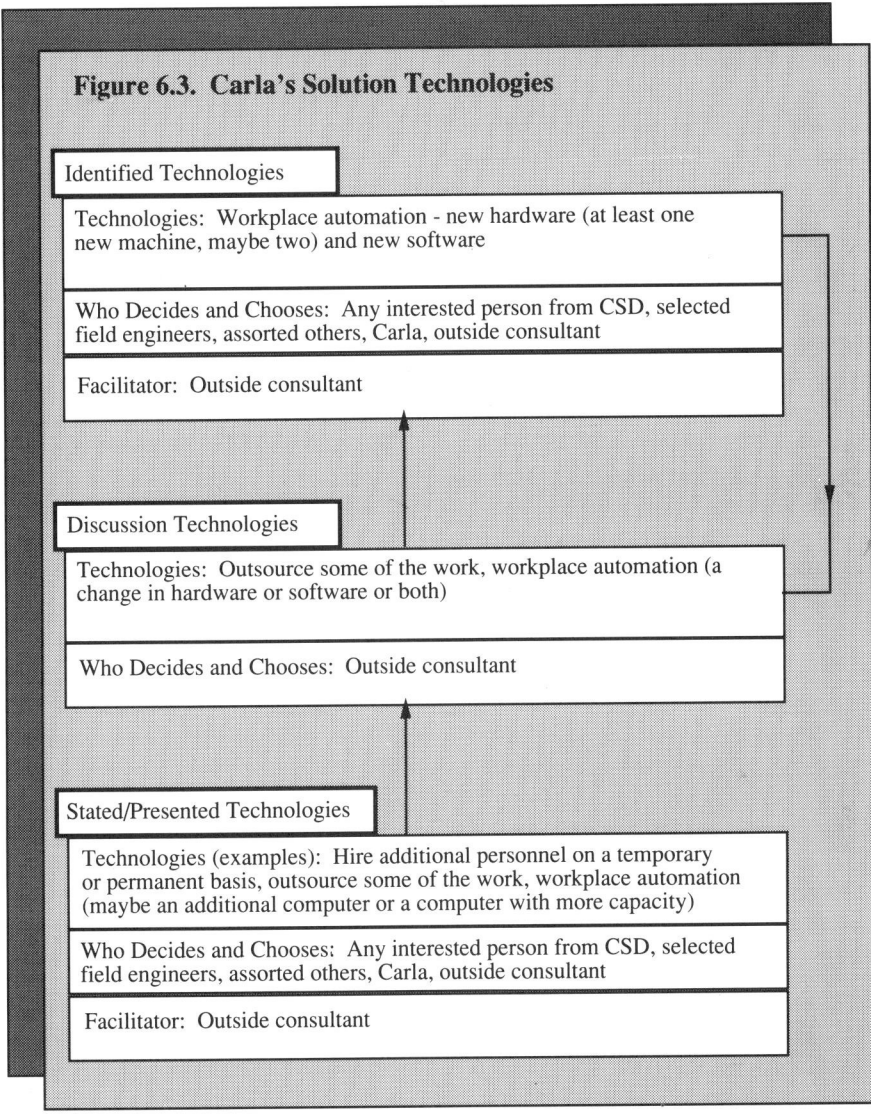

Figure 6.3. Carla's Solution Technologies

Identified Technologies

Technologies: Workplace automation - new hardware (at least one new machine, maybe two) and new software

Who Decides and Chooses: Any interested person from CSD, selected field engineers, assorted others, Carla, outside consultant

Facilitator: Outside consultant

Discussion Technologies

Technologies: Outsource some of the work, workplace automation (a change in hardware or software or both)

Who Decides and Chooses: Outside consultant

Stated/Presented Technologies

Technologies (examples): Hire additional personnel on a temporary or permanent basis, outsource some of the work, workplace automation (maybe an additional computer or a computer with more capacity)

Who Decides and Chooses: Any interested person from CSD, selected field engineers, assorted others, Carla, outside consultant

Facilitator: Outside consultant

thinking and decision skills necessary to complete a workplace automation project successfully. It is usually better to use retrained and additionally trained employees to do many necessary and optional computer tasks (or any other improvement task) than to terminate employees and hire others with computer skills who must learn the company's practices, procedures, and philosophy.

Well, both Chuck and Carla have decided to stay with current personnel — an important conclusion. But each organization as a whole has come to another important conclusion. Each organization has come to believe that it can "pull off" the project. The improvement project might just work after all. The realization/belief that you can successfully complete an automation technology transfer project is very important. The biggest or first barrier to successful technology transfer is the belief that you cannot handle the technology. If you do not believe that you can learn mathematics, you never will. If you do not believe that you can learn to use a computer, you won't. If you do not believe that you can integrate computing into your organization, it won't happen. Belief in your ability to master a skill is a powerful motivator/incentive to continue to work on those problems and difficulties that always arise during technology transfer. People who do not believe in their ability to solve such problems usually "give up" too quickly. Having decided on the technologies, Chuck and Carla must now decide if needed products are available. Since both Chuck and Carla have decided to use computing technology, automation products are considered and evaluated in the next chapter.

References

6.1: Watson, Bruce, "For a While, the Luddites Had a Smashing Success," *Smithsonian,* December 1993, pages 140–154.

Chapter 7
Stage Five:
Products Plain and Fancy

Unquestionably the popular thing in this world is novelty.

Mark Twain in *The Connecticut Yankee in King Arthur's Court,* Chapter XXXIX, The Yankee's Fight with the Knights.

During Stage Five a company decides which automation products, if any, can be used to implement the identified solution. A company also chooses product vendors and purchases the products. The employees undergo initial training in use of the products. It is a busy stage. The primary thing to remember at this stage is not to be swayed by product "novelty" — no matter how popular. You are buying these products for business purposes — to help you get work done. They are not toys and they are not entertainment. You need to make sure that you are buying them because they will do the work, not because they are popular or fun. The succesful completion of this stage requires that many questions be answered (26 to be exact). Do not be intimidated by the number, however, because if you have ever made a major purchase of any kind you are totally familiar with these questions and the process of answering them. The questions that must be answered during this stage are:

Stage Five Questions:

1. Who are the relevant people for this stage?

2. What are the specific tasks expected of chosen products?

3. Did Question 2 provide any information that warrants reevaluation of prior decisions?

4. What products are available that can complete these specific tasks?

5. Did Question 4 provide any information that warrants reevaluation of prior decisions?

6. What are the products for discussion?

7. What are the identified products?

8. Did Question 7 provide any information that warrants reevaluation of prior decisions?

9. What is the cost of each identified product?

(continued on the next panel)

Stage Five Questions (continued):

10. Did Question 9 provide any information that warrants reevaluation of prior decisions?

11. Which products, if any, are affordable?

12. Did Question 11 provide any information that warrants reevaluation of prior decisions?

13. Which products, if any, will be purchased at this time?

14. What are the available product vendors?

15. Did Question 14 provide any information that warrants reevaluation of prior decisions?

16. What are the similarities and differences between vendors?

17. Did Question 16 provide any information that warrants reevaluation of prior decisions?

18. Who are the vendors for discussion?

19. Who are the identified vendors?

20. Did Question 19 provide any information that warrants reevaluation of prior decisions?

21. Who will do product installation?

22. Did Question 21 provide any information that warrants reevaluation of prior decisions?

(continued on the next panel)

Stage Five Questions (continued):

23. How will initial training be completed?

24. Did Question 23 provide any information that warrants reevaluation of prior decisions?

25. What is the budget and schedule for the rest of the project?

26. Did Question 25 provide any information that warrants reevaluation of prior decisions?

Because this process should be familiar to anyone who has made a major purchase and because it is so similar to prior stages, we will not discuss the specific questions in great detail. We will only highlight a few points.

How to Find Product Information

It would be nice if we could, at this point in the book, just tell you which products you should buy. Unfortunately, we cannot do that. Nobody can. No one computer product is right for all people in all situations. But this is true of all purchases: No one financial investment is right for all people in all situations, no one car is right for all people in all situations. Nor are we going to do an exhaustive discussion of the many computer products being sold today. Why not? Because this information already exists in a variety of places. Also because an adequate discussion would make the book several thousand pages long. What we will mention at this point is how to find the product information that will help you and the other members of your organization decide which products the organization should buy. Fortunately, you gather information on possible computer products for use in much the same way you gather information on any other product (e.g., financial investments, a car). This method is basically the "hard work and doing your homework" method (see Figure 7.1). You can gather much valuable information by talking to vendors that sell computer products and to people currently using workplace automation.

Figure 7.1 How to Find Specific Product Information

Read Computer Magazines
and Newspapers:

Examples:

    - BYTE
    - Datamation
    - Infoworld
    - PC Magazine

Also noncomputer ones.
  Example:

    - Consumer Reports

Why Read (examples):
Ratings of hardware and
software and criteria used,
how to solve problems,
advertisements

Read Books about Computers
(examples of topics):

- How to use a specific
  software package
- How to buy a personal
  computer
- How a computer works

Where to Find Computer Magazines,
Newspapers (examples):

- Local bookstore
- Local library
- Local office supply store
- Local paper and magazine store
- Local computer store

Where to Buy Hardware and
Software Products (examples):

- Local computer store
- Local office supply store
- Local electronics store

Basic Types of Hardware and
Software Vendors:

- Local full-service
- Local discount
- Mail order (varying levels of
  customer service provided)

Where Computer Courses are Offered
(examples):

- Local adult community education program
- Local college's continuing education
  program
- Local computer store

Talk to:

  - Friends
  - Colleagues
  - Computer
    salespeople

Check the Yellow Pages under data communications,
data processing and computers, attend computer
shows, fairs and expositions

Those in your industry who are already using automation are an excellent source of information. Maybe there are people in your industry or a related one who regularly meet to discuss computers and automation. Ask your colleagues. Find out if a product that interests you has a users group, which is a group of people using the same product (usually the same software). They meet to trade information (e.g., what they have found out about the software, how they use the software). Ask your colleagues; ask your local product vendor if such a users group exists. Many computer books exist that discuss specific products (e.g., how to use a personal computer, how to use a particular word processing package). You can also read some of the many computer magazines such as *BYTE, PC Magazine, Microcomputing, Dr. Dobbs Journal, Datamation, and Infoworld.* These magazines always have product advertisements and usually have product reviews. *Consumer Reports* sometimes reviews and rates automation products. You may also wish to attend computer shows and fairs or join one of the many computer associations such as the Association for Computing Machinery (ACM) or American Association for Artificial Intelligence (AAAI). Most computer organizations include literature and magazines as part of the membership. Read the literature even if you do not understand all the information. Many of these organizations also hold conferences in nice places. You may wish to attend some of the conferences.

This "hard work and homework" process allows you to narrow your choices and then closely examine a few. It is unlikely that you will find the perfect computer product. Hardly any product of any kind (short of a completely customized product) is "made just for you." Some adaptations to the computer products may have to be made (just the way you need to make adaptations to any other mass-produced product such as a new car). This does not mean, however, that you choose a product that requires major adaptations on your part. If the search for computer products does not find any suitable products, then a company may have to make do with current technologies and products until — if and when — a suitable product is developed. A company should never purchase an unsuitable product just because it is the only one around. The more unsuitable the product, the less likely it is that the technology represented by the product will be successfully transferred. The lack of a suitable product (especially the lack of a suitable software product) may tempt some people to consider development of a customized software product (See Chapters 9 and 11 for further discussion of customized versus mass-produced software products). If possible, do not enter the world of customized computer applications. This world can be hell. If you do enter it, make sure that you know who makes and owns the customized package. You need to be especially careful if you are using outside consultants. You do not want to be in the position of having paid their development costs. That is, you paid them to write a package for you that they then sell to other companies and you get nothing from their sales.

Automation Product Categories

When evaluating automation products, three categories of products must be evaluated independently and in order. The three categories are software, hardware, and vendor. Software is chosen first because hardware should be chosen based on software requirements. The hardware under consideration should be hardware that runs or works with the chosen software. All software does not run on all hardware. If you choose the hardware first, you run the risk of not being able to operate the needed software on the chosen hardware. A hardware requirement is, therefore, that it run or work with chosen software. After you choose the software and hardware, you choose vendors from whom to purchase the selected products. If you choose the vendor first, you usually restrict your software and hardware considerations to only those sold by the chosen vendor. This vendor may not sell the software and hardware most beneficial for you. You should also know that your hardware and software do not have to be purchased from the same vendor (see the Purchase Guidelines listed in Chapter 9). You also do not need to purchase all your software or hardware from one vendor.

Avoid Spillover

When you choose a product you should also make sure that use of the product does not result in *spillover*. Spillover occurs when the effect of automation has been allowed to spill over into areas that were not intended to be affected by the automation project. Automation of one task or a set of tasks can change the way other tasks are done. Here is an example. Suppose that an office of medical professionals (or social service professionals or legal professionals or others) has installed a centralized, automated desk and appointment calendar for each professional. Suppose this was done to allow one person to schedule client appointments for all the professionals and to make it easier to produce monthly reports on the number of clients seen by each professional during the month. Suppose also that each professional is supposed to use the calendar to record activity unknown to the person scheduling client appointments (e.g., meetings, consultations with other professionals, lunch). Suppose also that the professionals spend more hours using the automated calendar than they had previously spent using their nonautomated calendar of activity. The automated calendar is time-consuming and difficult to use. The recording and manipulation of calendar information are administrative tasks for this organization. Any increase in time spent by the professional staff in administrative duties decreases the time they can spend in direct client contact. Direct client contact (direct contact with consumers) is the mechanism through which the organization's product is produced (activities occurring during direct client contact are production tasks). A decrease in the time available for production tasks decreases the number or quality of the organization's products. In this example the effect of automating an administrative task has been allowed to spill over into an unintended area of automation effect — the production area. On the whole, organizational productivity has been reduced.

Consider also an automated case record system that Chuck would like to find and purchase for his organization. There are two basic views of a case record system (automated or not): the social worker's view and the organizational administrative view (i.e., Chuck's). Clearly the two perspectives are interrelated, but the perspective chosen changes the functioning of the automated system. An automated system designed from the social worker's perspective is one designed to enhance product development. The product of this social service organization (sometimes known as a family agency) is change in a client's behavior, attitude, or feelings. The product is produced through the interaction between a therapist and the client. Family agencies are, therefore, information processing agencies. Information processing mechanisms are the tools of the organization. A case record system developed from the product development perspective should not only quantify case information, but also facilitate the client-therapist interaction in a way that enhances product quality. Information necessary for quality product development must be addressed first. Once this is done administrative reporting requirements can be discussed. The reporting information should be obtained from the case record information. Any "extra work" required by the difference between therapist needs (product development needs) and reporting needs is done on the administrative side of the company.

An alternative method is to downplay product development needs and see the administrative reporting requirements as primary. When software is developed from this perspective, one often sees pages of the reporting forms presented as screens for the therapists to complete on each case. This approach is often flawed because reporting requirements change frequently and this software hinders service delivery. Service delivery is hindered because such software usually results in an increased workload for those people directly involved in product development. Product personnel often begin to keep two sets of records and complete two sets of tasks. They maintain information useful for product development and information on the same case useful for administrative reporting. The product development (case) notes are usually handwritten, while the information useful for reporting is either entered on a paper form (and later entered into a computer) or directly into a computer using a screen form. Therapists (product development personnel) begin to do both product development and administrative (reporting) tasks. Another problem with this approach is that reporting requirements are not always designed in the optimal format for product development. The variables (answers to questions) required on administrative reports often become the standards against which the product is measured. Such reporting variables are not always the best variables for measuring product quality yet there are no other easily obtained variables to use. The development of good reporting requirements is similar to the generation of a good survey. It requires skill. Some reporting requirements we have seen are good, but many are bad. Good or bad, however, they must be met. They must not be met, however, by harming other essential parts or organizational functioning.

After Purchase and Installation

This stage also includes product installation. Automation installation means connecting all computer parts, plugging the computer into electrical power, configuring computer networks and communication equipment, and getting needed software ready to use. You should install your equipment as soon as it arrives. Many automation products have warranties that expire in 30, 60, or 90 days. All products should undergo regular use during most of the warranty period because you want to find any "regular use" defects while the warranty is valid. If computer equipment is going to malfunction, it will usually malfunction immediately. A mistake that many people make is to buy the products and let them sit in a box for a long period of time. The first time they use the products, they break, but since the warranty is expired, there is little they can do about it. Also included in this stage is use of any training material provided with the product.

Once the products are purchased, installed, and understood (at least to the level of initial training), it is time to use them to do daily activities. This task may require much time and double work for a company. One time-consuming process for a company is translating daily activities into commands that a computer can use to assist employees. This translation process often requires modification of existing information and policies (as shown in the earlier discussion of absence and accrual time automation) and always requires the entry of existing or modified data into the computer.

After translation, evaluation of the accuracy of the translation is necessary — another time-consuming task. A common way to evaluate the accuracy of automation instructions for a dynamic task (e.g., payroll; payroll amounts and employees often change from payroll to payroll) is to run in parallel. Running in parallel for dynamic tasks means that the new automated system operates concurrently with the manual one with each doing the same job. An automated payroll system, for example, operates in parallel with a manual one when each system generates the necessary reports and payroll checks (usually one set of checks is printed on plain paper rather than a check form). The two systems continue in parallel until the user (person using the old and new system) is sure that the new one works accurately. Clearly for such parallel runs to be consistent, everything entered or used for the manual system must be entered and used by the automated one — hence the double work. Much of the work is also often done repeatedly since the system is usually tested in parts rather than all at once (e.g., a few employees rather than all are entered into the automated payroll system) and only when one part tests well are others added and tested.

Accuracy evaluation is done in the parallel system by checking both the manual and automated output. The manual payroll reports and checks, for example, must be compared to the automated ones. If they are not the same, then resources must be spent to find the reason for the difference and to make necessary changes. Accuracy evaluation for static tasks (e.g., form letters) is usually done by checking the automated output with the existing manual output. A form letter created using word processing, for example, is compared to an existing form letter that was created using a typewriter. For such static tasks, there is no parallel run — the word processing letter and the typed letter are not created at the same time and then compared. As with the evaluation of dynamic tasks, however, if the automated and typed letter are not the same, then resources must be spent to find the reason for the difference and to make necessary changes. The amount of resources needed for this step depends on the size and complexity of the project. The larger and more complex the project, the more resources are required since there is more data to enter and more checking to do. You should also realize that technology purchases (like many other purchases) have a habit of living longer (or shorter) than one often expects. Expect to use any software at least five years. Five years can be a very, very long time to live with a mistake. So buy products carefully.

Data/Information Responsibility

Another thing that must be finalized before the automation goes into regular use is who has responsibility for what. It is important to have at least two people familiar with all automation tasks. This allows one person to be sick or go on vacation and also keeps you from being blackmailed. Automation blackmail occurs when the only employee who knows how to do key automation tasks continually asks for salary increases. You should decide now — if you have not already — who the responsible people and their backups are. It is important to confirm various aspects of automation responsibility before "going live." The importance of responsibility in using automation cannot be overemphasized because if computers are to have the proper impact people must be responsible. It is the responsibility of individuals to use the most productive aspects of a product in the most productive way and to take their responsibilities seriously.

Computers allow human beings to abdicate their responsibilities as few technologies can. Computers seem to be used by some people to escape from all responsibility for an action. How many times have you heard "I am sorry. The computer is down. The computer has made a mistake. I cannot help it if the computer says that you are dead." The computer is replacing "It's not my fault. Someone else did it." Computers are easier to blame because they do not fight back when falsely accused the way humans usually do. We see this as a serious problem because there seem to be many people who want privileges and rights, but not responsibilities. The solution to this problem is not to do away with computers, but to educate and encourage people to take responsibility for any task in which they are involved. If there is an error with a computer, it is some human's responsibility and it should be fixed. Abdication of human responsibility is maddening when you are trying to correct an error in your charge account, but it can be devastating in other applications (e.g., weapons systems, nuclear power plants).

Automation responsibility also includes data accuracy. Make sure that data are entered correctly and that all necessary data are available. Part of ensuring this is to make someone responsible for it. Make someone responsible for developing, enforcing, and executing procedures that ensure the accuracy of information in and from your computer. You should also make someone responsible for developing, enforcing, and executing procedures to make corrections or updates to information in and from your computer. There should be a person responsible for making corrections to existing data or output when errors are reported. There should be a structured way to report an error to the person responsible for making corrections. Finding and reporting errors implies that everyone is looking for them. Always look for errors. Manipulate and report data in a way that makes errors easy to find. No matter how careful you are with data entry and output, mistakes will happen. The more careful you are with data entry and output methods, the fewer mistakes, but you will never eliminate mistakes. Everyone in the company has a responsibility to look for problems and to report them to someone responsible for tracking down and fixing the problem. No one should assume that all information from a computer is correct. It is true that a new notion of quality is to design systems so that errors cannot occur. This is an admirable strategy, but often impossible to realize. Consider the problem of manually entering inventory counts: the count, the entry, etc.

You should also make someone responsible for using any additional information on computer that falls outside the daily computing tasks. Unless you use the information on your computer, there is no sense having it there. Automation usually increases the amount of data easily available for use. These data should be used. Responsible use of this information does not mean requesting an infinite number of reports or expecting that any piece of information or report you want can be generated in seconds. Responsible use of your information means having a reason for each report request and using the information in that report to contribute to or improve company functioning. Running many reports and letting them pile up without ever looking at them is not responsible use. Use the information that you generate.

Last, but definitely not least, you should maintain decision responsibility. Do not turn your decision-making authority and responsibility over to a computer. It should be clear by now that automation should assist a human being in making a decision (e.g., provide needed information), but should never make a decision. The human mind is much smarter than a computer and a "mind is a terrible thing to waste." Yet as we mentioned at the beginning of this discussion of responsibility guidelines, we feel that some people would rather waste their minds than take responsibility for a decision. The more responsibility you abdicate to a computer, however, the more likely it is that something will go wrong. Productive computer use requires increased personal responsibility, not decreased. Decreasing your personal decision-making responsibility will lead to disaster for you and others affected by your automation.

Chuck, Carla, and Automation Products

Chuck and other members of his organization decided to examine a variety of software that might benefit the social service organization even though they knew they could not afford to purchase much. They looked, for example, at various billing packages. None of the packages, however, contained all of the features needed by the social service organization. None, for example, allowed multiple people to be scheduled with the same professional at the same time. Such a situation is common in a social service organization (e.g., when all members of a family are seen at the same time). The billing software would not allow a bill to be generated for anyone who did not have an appointment. There also seemed to be no reasonable way to generate and record a sliding fee scale charge. There was a complicated way to "trick" some of the software packages into billing on a sliding fee scale, but the "trick" was really too cumbersome to use. They decided against purchasing any billing software at this time. They also looked at case record software, but again found nothing they thought would work for this social service organization.

Chuck and the others finally decided that the only realistic software purchases that could be made at this time were word processing, spreadsheets, a database manager, or statistical software. Chuck again checked the budget to see what he could afford to spend at this time. Because of some lucky breaks, he had more money to spend on automation than he thought he would have. He did not, however, have enough money to purchase all four software package possibilities. The organization finally purchased two personal computers, two printers, two copies of a specific word processing package, and two copies of a specific spreadsheet software package (see Figure 7.2). Susan (the secretary with a fair amount of computer experience) played a very active role in deciding and choosing software and hardware. She also spent much time explaining various computer products to her colleagues and showing them how to use them. Almost everyone in the organization was very enthusiastic about the entire product selection process and found it all very interesting. Mary (the social worker who has been against computers from the start) was the only one uninterested in the entire project. Mary still thought the whole automation idea was silly and refused to go along on any of the field trips to look at automation products.

During this stage Carla and others within her organization were also busy examining products for possible use. They first searched for mass-produced (off-the-shelf) payroll and warehousing software that would meet their needs. They found none. That led them to the conclusion that customized software (written in-house) was the only viable solution for the company. Carla, working with the Computing Services Department, hired programmers and systems analysts to work on the customized software project. Some of these new people are to take over tasks done by current CSD personnel so that they can write the customized software. It was difficult to find people who were right for this project, but they did. Carla and others also evaluated various machines produced by more than one manufacturer. It was not clear to anyone in the company that they wanted to buy another machine from the vendor who sold them the "nightmare" machine. In the end, however, four machines were purchased from this same vendor (see Figure 7.3). The vendor was as anxious as anyone to end this nightmare and offered Carla a very good deal on the machines — and made sure they were the right ones for the job. Two machines will run the warehousing software and be located in a warehouse on the docks. The other two will be in the current CSD location and be used to run payroll. All machines will be kept current with each other and serve as backup machines for the others. This is accomplished using a shared file system residing on one machine and shared with the others (such a strategy actually provides $n + 1$ redundancy).

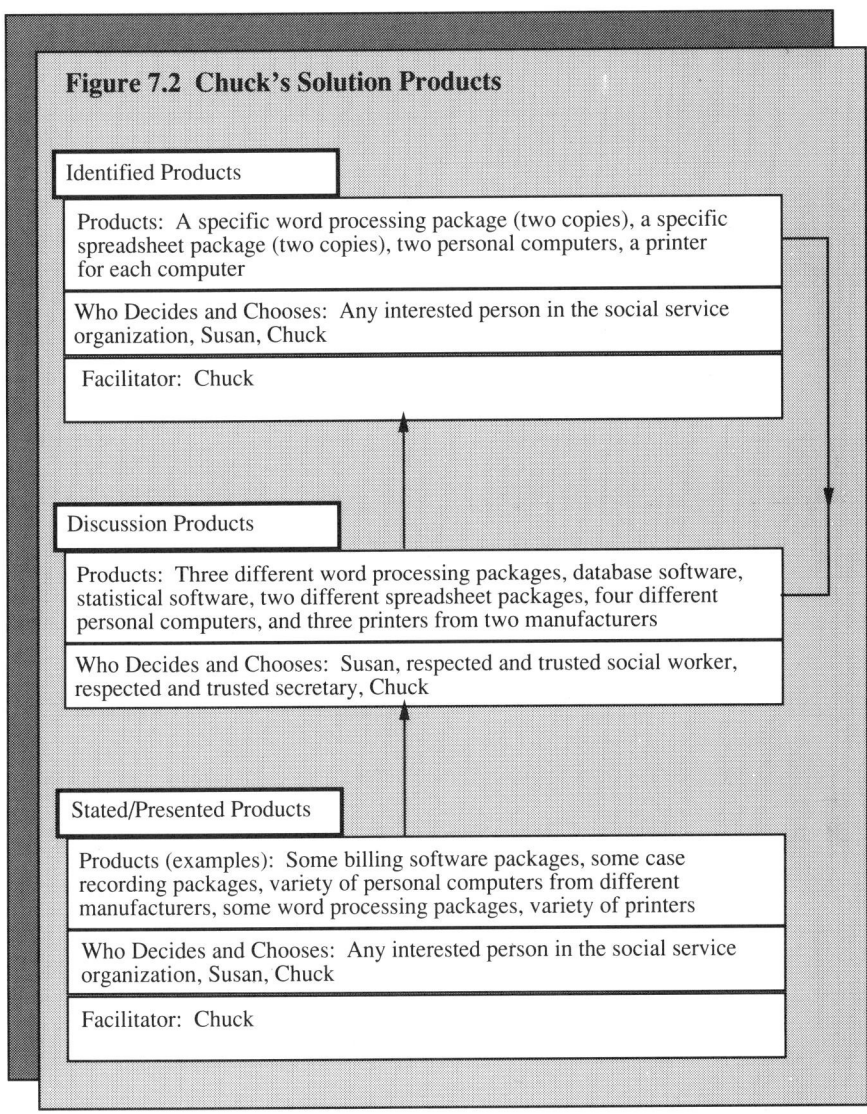

Figure 7.2 Chuck's Solution Products

Identified Products

Products: A specific word processing package (two copies), a specific spreadsheet package (two copies), two personal computers, a printer for each computer

Who Decides and Chooses: Any interested person in the social service organization, Susan, Chuck

Facilitator: Chuck

Discussion Products

Products: Three different word processing packages, database software, statistical software, two different spreadsheet packages, four different personal computers, and three printers from two manufacturers

Who Decides and Chooses: Susan, respected and trusted social worker, respected and trusted secretary, Chuck

Stated/Presented Products

Products (examples): Some billing software packages, some case recording packages, variety of personal computers from different manufacturers, some word processing packages, variety of printers

Who Decides and Chooses: Any interested person in the social service organization, Susan, Chuck

Facilitator: Chuck

Carla's purchase of four machines instead of one (or two) illustrates more than a good business deal. It illustrates the general downsizing of the computer industry (not of the personnel kind — the machine kind). Machines are getting smaller, more powerful, and cheaper. The movement is away from the "one big mainframe" technique that Carla had been using [7.1]. While waiting for machine delivery, Carla completed construction on the machine room that will house the warehousing machines. She modified the current machine room to house the new payroll machines alongside the current (soon to be "old") machine.

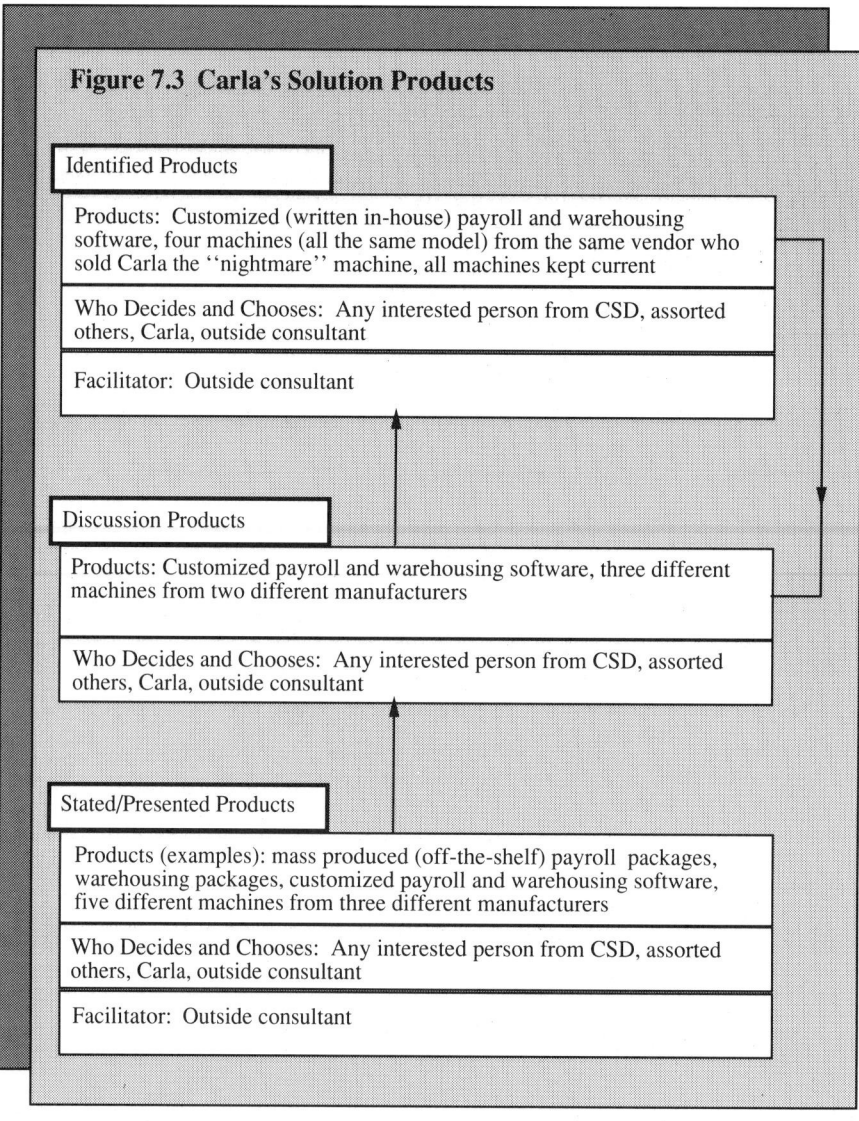

Figure 7.3 Carla's Solution Products

Identified Products

Products: Customized (written in-house) payroll and warehousing software, four machines (all the same model) from the same vendor who sold Carla the "nightmare" machine, all machines kept current

Who Decides and Chooses: Any interested person from CSD, assorted others, Carla, outside consultant

Facilitator: Outside consultant

Discussion Products

Products: Customized payroll and warehousing software, three different machines from two different manufacturers

Who Decides and Chooses: Any interested person from CSD, assorted others, Carla, outside consultant

Stated/Presented Products

Products (examples): mass produced (off-the-shelf) payroll packages, warehousing packages, customized payroll and warehousing software, five different machines from three different manufacturers

Who Decides and Chooses: Any interested person from CSD, assorted others, Carla, outside consultant

Facilitator: Outside consultant

When the new machines arrived, they were installed and CSD got to work both writing the new software and moving the old software to the new machines. They had to move some of the old software over because it would be impossible to have all the new software written and ready to go by the time the old machine was supposed to be removed (when its maintenance contract expired). There were, of course, many additional hardware problems and many tense moments. Nonetheless, the project marched on to the parallel runs in which the new machines and the old machine run the same programs at the same time. If they all come up with the same "answer," everything is fine. If they do not, it is essentially "back to the drawing board." After many weeks of parallel runs, everything was finally ready to go. The old machine was removed on time. (Isn't it amazing how months of hard grueling work can be reduced to just a few sentences?)

References

7.1: Miller, Michael J., "Networked PCs: Nothing Personal?" *PC Magazine,* Volume 13, Number 1, January 11, 1994, pages 77–78.

Chapter 8
Stage Six:
Let the Facts Reign

Well, it rained mortar and masonry the rest of the week. This was the report; but probably the facts would have modified it.

Mark Twain in *The Connecticut Yankee in King Arthur's Court*, Chapter VII, Merlin's Tower.

"Fallen Statue, Surrey, England," photo © copyright Christi Carter, 1993.

In this last stage of the technology transfer process, a company begins to use the chosen technology (in this case, workplace computing) in daily workplace life. The organization ends the improvement project and determines its effectiveness. Was the project worth all the effort? Was the project successful? This is clearly not a small point. In determining success, however, it is sometimes necessary to modify the project reports with facts. In improvement project reports, there seems to be a general tendency toward either glowing reports of success (Everything is great. Everything is wonderful.) or gloomy reports of failure (Everything is wrong. Everything was better before.) All in all, however, this last stage should be fairly tranquil and anticlimactic. There are only three major questions that must be answered in Stage Six.

Stage Six Questions:

1. Who are the relevant people for this stage?

2. When, exactly, will the project end?

3. How effective/successful was the project?

Stage Six opens for Chuck when he and the others start using the chosen computing to do "live" daily workplace tasks (e.g., word process a "real" letter, produce a "real" budget spreadsheet). This stage opens for Carla right after she disconnects the old malfunctioning machine and relies only on the four new machines for daily work. For the Computing Services Department personnel, passage into this stage means a chance for a normal life and job; 40 to 50 hours at work each week, no more sleeping on the floor of their offices, taking a day off — maybe even a vacation, only occasional use of antacids, and maybe even an exercise program to lose some of the weight gained during the last couple of years.

The relevant people for this stage change with the question being asked. For the second question, the relevant people are the project facilitator(s), the person(s) responsible for the evaluation, those responsible for writing the report, and those to whom the report will be delivered. To answer the second question, three different project end points need to be determined:

1. When does the initial adjustment period end?

2. When does the evaluation period end?

3. When is the final report on project success due?

Despite all the training and all the planning, there are always a few adjustments to make after a project goes live. Dress rehearsal is not the real show. But if you have followed the technology transfer process described in this book, these adjustments should not be major. Nonetheless, you should allow some adjustment time before evaluating the project's success. How much time? One month or so. There is no need to prolong this stage. The facilitator should, however, confirm with those in daily direct contact with the computing that no unusual circumstances exist that warrant a longer adjustment period. It is not fair (and often not possible because of time constraints) to evaluate a project while adjustments are still being made.

How long should the evaluation period be? No longer than a few weeks. Again, there is no need to prolong this stage. The content, form, and due date of the final report is decided by the facilitator, the person writing the report (if not the facilitator), and whomever the company usually involves in such decisions (e.g., Carla, Board of Trustees). A formal written report is not always necessary. Chuck, for example, will give a verbal report (with some handouts) to the Board of Trustees. A brief report will also be sent to the foundation that funded purchase of some of the computing equipment. Carla wants an extensive written report from the outside consultant. She also wants review and discussion of the report by CSD personnel before she accepts the report. After acceptance of the report, Carla's facilitator/consultant leaves and Chuck ends his facilitator role — for a couple of days at least. These technology transfer projects are finished.

The somewhat arbitrary determination of when to end the project recognizes two things about technologies:

1. The value of technology decisions depreciates over time (sometimes rapidly).

2. Success can only be determined with monotonic correctness (see below).

The longer you wait to determine project success, the more likely it is that technology decisions made during the project will depreciate greatly in value. Most technologies (especially high-tech ones) change fairly rapidly. It seems that the day after you buy a product, a better one comes out or is announced. That is just the way it is. Almost everyone feels some sense of regret (some sense of decision depreciation) almost as soon as a purchase is made. It is better to have these regrets occur after you have declared the project a success than to have them influence the determination of project success. The sooner you evaluate and end the project, the fewer should be the regrets influencing the evaluation. The rapidly changing nature of high-tech industries such as the computer industry also means that a company can get behind by being ahead.

The company or people who automate after you probably have more advanced equipment. Quite quickly, others have state-of-the-art equipment and you do not. When it comes to the use of rapidly changing technologies like computing, the first shall be last (at some point in time). But this does not mean that you should wait forever to automate. Waiting is often less productive than being "last." Monotonic correctness means that the determination of project success is really contextual. The decision about whether a project is successful or not reflects environmental conditions in effect at the moment that decision was made. If the environmental conditions change, the value of decisions made regarding a particular technology may change. The Titanic is a good example of this situation. The Titanic was judged a superb ship — a superb set of technologies — in calm seas before it hit the iceberg. When the environment changed, when the Titanic hit the iceberg, the Titanic and its technologies were no longer superb. They became a technological, political, and human disaster.

Just because a project is finished, however, does not mean that everything is frozen in time. As the project ends it is time to start looking and preparing for the next wave of technology. Everyone should continue to develop more expertise through continued use of chosen products. Adjustments should be made as necessary, including the upgrade of chosen products or the introduction of new ones. Automation products and any other products used by a company must be continually monitored and evaluated. As the products break, become worn, or become outdated, they must be repaired, replaced, or upgraded. Ending the project does not mean, therefore, that all progress and changes stop. Chuck, for example, comes to realize (after Susan points it out to him) that it is more effective for him to type his letters and memos directly into the computer himself. He had been writing the letters and memos longhand or dictating them to a secretary (as he always had). The secretary then word-processed the material, returned it to Chuck for editing/revisions, and then made the revisions. Susan's suggestion helped Chuck better understand how to use workplace automation productively. Susan's suggestion increased Chuck's computer literacy.

Computer Literacy

A successful workplace computing technology transfer project develops computer literacy. It also develops mechanisms to increase literacy after the project ends. People are able to help each other (as Susan did) and they are able to help themselves (e.g., they know how to find information that they need). We define computer literacy as knowing what you need to know about computers to do what you need to do [8.1–8.5]. Workplace computer literacy means knowing what you need to know about computers to do your job. A person with workplace computer literacy should be able to make informed decisions about the details of productive, daily use of workplace computing. A computer literate person does not need to know all the answers, to have all the information, but should be able to generate possible solutions to any automation problem that arises. The computer literate person should be able to determine the information required to solve a typical computing problem and the possible sources of required information. Computer literacy also means being able to recognize important information when you see it. When company personnel are computer literate, they are self-sufficient users of automation products and automation technology. Self-sufficiency is required for successful automation technology transfer because daily life experiences are so varied and detailed that it is impossible to teach someone the proper response to each possible daily situation. No amount of training and instructions is detailed enough to cover all possible problems that may be encountered during a workday.

To be self-sufficient, employees do not need to know everything about both the organization and the computers. The employee must know how to do his/her job and the characteristics of automation that may make the job easier, more efficient, and more effective. There will always be something new to learn, but the self-sufficient employee is able to integrate use of the computer into the daily work process. The self-sufficient employee is always looking for new ways to make the computer an extension of the work situation. In addition, this employee is aware that such a process is a fundamental aspect of the job. The automation self-sufficient employee is aware that making decisions about computers and the job is routine, ordinary, and expected. Such decisions are as expected as decisions about applications of the photocopy machine, telephone, typewriter, and other office equipment. If an employee is a self-sufficient worker (and reasonably self-confident), he/she can be a self-sufficient user of automation.

The pieces of information that all people need to be workplace computer literate cannot be precisely specified because the number and variety of information pieces required depend on the tasks to be completed. The number and variety depend on a person's workplace literacy peer group and what the person needs to know to do a job effectively. A person writing word processing software, for example, must have an in-depth knowledge of the best programming languages and methods for producing word processing software. This person should also know when and how word processing software is superior to writing with a typewriter or pen. A person selling computers requires some word processing literacy, but not to the same depth or degree as a person who writes word processing software. The salesperson requires a computer word processing sales literacy (in addition to a general computer sales literacy). Computer sales literacy usually involves knowing the range of computer equipment, the range of software, the competitor's stock of equipment and software, and the relationship between a computer's capabilities and what the potential customer wants to do. The literacy peer group for the professional who creates word processing software contains other professionals who do the same thing. The literacy peer group for the computer salesperson contains other computer salespeople. Your workplace computer peer group contains members of your industry. Within each peer group there is usually a large range of competencies.

So how long does it take to learn about computers and workplace automation? The answer to this question is both 15 minutes and the rest of your life. It takes about 15 minutes for the average person to understand basic computer concepts. In 15 minutes you can be well on your way to a basic computer literacy. It will take the rest of your life to become truly intuitive and knowledgeable about computers. With computers, as with all other knowledge areas, the acquisition of one piece of information usually makes you aware of how much you still have to learn. The amount of time it takes you to become computer literate depends on the depth of computer intuition and knowledge required. The depth to which you learn about computers depends on the degree to which you want and need to know computers. The attainment of any level of computer literacy requires time, patience, determination, and effort. The more computer literate or competent you wish to become, the more time and effort you need to put into the project. But the attainment of computer literacy is no harder and no more time-consuming than the attainment of any other literacy or competency. We believe that the best way to become computer literate is through experience making computer-related decisions and using computers. The technology transfer stages listed in this book provide such experience.

Measuring Project Success

So what did Chuck and Carla decide? Were their projects a success? How, specifically, do Chuck and Carla (and anyone else) determine project success/effectiveness? Basically a successful project is one that does what the organization thought it would. To determine project success, a company needs to ask whether its expectations were met. If yes, the project was successful. Some companies state their expectations in terms of extensive quantitative benchmarks. Others do not. To determine project success, five questions must be answered:

1. What did you expect the project to accomplish?

2. How will you measure these expectations?

3. To what extent were these expectations met?

4. Did anything unexpected happen?

5. All things considered, was the project worth the effort?

Why wait until now to state clearly the expectations of computer use that determine project success? — Because expectations often change over the course of the project. A major reason for such change is that some of the original expectations may not have been realistic. The technology transfer process described in this book is designed to bring such unrealistic expectations into line with reality. Chuck, for example, originally expected to automate everything in the social service organization — quickly and easily. The original expectations were not very realistic. Chuck did not understand enough about computers to know his expectations were unrealistic. As the project progressed, however, his understanding increased and his expectations changed. His expectations became more realistic. It was these final, more realistic expectations that were met (see Figure 8.1). Carla originally thought that her computer problem could also be solved quickly and easily. It turned out, however, that her computer problem was much more extensive and difficult to solve than she originally thought (see Figure 8.2).

Figure 8.1 Chuck's Project Expectations and Success

Original Expectations: Automate everything (e.g., reports, budgets, case records), this can be done quickly and easily

Final Expectations: Use automation to generate reports, budget sheets, letters, and memos; automation takes much time and effort

Original Expectations Met: No

Final Expectations Met: Yes

Surprise Project Results: Mary uses computers in therapy, overall organizational computer literacy achieved, most people see the organization in new ways (e.g., information a major component)

By and Large, Was the Project Worth It? Is the Project a Success?:

YES

Besides stating final expectations and whether they were met, a company also needs to determine if anything unexpected happened on the project. Even "unsuccessful" projects (ones where final expectations were not met) may have unexpected benefits that outweigh the lack of success. The unexpected benefits may make the project worth the effort after all. Consider some vacations you might have had. Maybe what you expected to happen did not (e.g., the flight was cancelled, the hotel was a dump), but it was a fine or great life experience anyway (e.g., you met some wonderful people, you saw a bear). Consider college. A reasonable expectation for any semester is that you will learn more about a specific subject (e.g., mathematics). But there are also many unexpected benefits of attending (e.g., increased self-sufficiency, self-confidence, interpersonal skills, ability to deal with people in authority). So maybe you flunked the math test, but maybe the semester wasn't a total waste after all. It does not mean, however, that you should not go back and take the mathematics course again. There were many unexpected benefits on both Chuck's project and Carla's project.

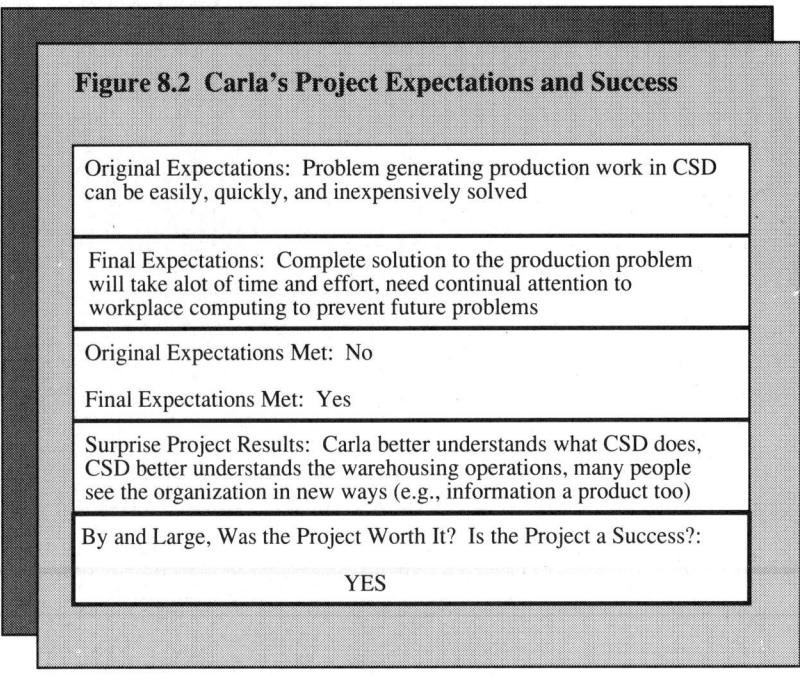

Figure 8.2 Carla's Project Expectations and Success

Original Expectations: Problem generating production work in CSD can be easily, quickly, and inexpensively solved

Final Expectations: Complete solution to the production problem will take alot of time and effort, need continual attention to workplace computing to prevent future problems

Original Expectations Met: No

Final Expectations Met: Yes

Surprise Project Results: Carla better understands what CSD does, CSD better understands the warehousing operations, many people see the organization in new ways (e.g., information a product too)

By and Large, Was the Project Worth It? Is the Project a Success?:

YES

Chuck's project had one very unexpected benefit. Mary (the social worker who had been against computer use from the beginning) embraced computing. Yes, Mary. Well, actually, Mary still thought computers were uninteresting and wasteful for administrative use, but she did find two innovative ways to use computers as part of the therapy process with children and adolescents. One use is as an incentive for adolescents to keep their appointments. Keep your appointment and you get to spend time learning about computers. With this incentive, the number of adolescent no-shows dropped. Mary also started using computers as a tool in the "Storytelling" technique she had long used with multiproblem, hard-to-reach children and adolescents. In this technique, Mary and the adolescent or child take turns starting and completing sentences. This technique gathers information about the client and develops the practitioner-client relationship. Before computers, this was done either verbally, with pad and pencil, or with a typewriter. For some reason, using a computer with word processing is more effective. One reason that Mary was able to see a use for computers in therapy was because she was involved in the mutual problem-solving and shared responsibility groups from the beginning. She had always been a "relevant" person even though she had objections and chose to participate in the process only marginally. This involvement in the process (though limited) made her aware of computer capabilities and allowed her to find uses for them.

Technology Transfer Process Retools the Organization

An unexpected benefit found in both projects is that the people within each company came to a better understanding of what others in the company do. This understanding developed largely through use of the mutual problem-solving and shared responsibility technique. Many job functions in many companies are done in a fair amount of isolation. You know what you do, but you do not really have a clue as to what anyone else does. The mutual problem-solving and shared responsibility technique forced people who normally do not talk to each other about their work to talk in depth to each other about their respective jobs. Another common unexpected benefit is that the people involved in each project came to view their respective organizations in new ways. The evaluation of general solutions and solution technologies developed new views of the organization (e.g., how important a particular department or organizational task is). The evaluation of workplace computing technology developed a view of information within the organization that most people did not have. Almost everyone in each organization came to view information itself as a major component of the business. Carla, for example, became very interested in the nature and flow of information in her organization. She had always thought of her business as being fairly physical; hers was the business of lifting, moving, sorting, and collecting goods. She now sees that information about those goods is just as important as moving them around. This new view leads Carla to consider the formal addition of information to most of the organizational tasks and products. Carla wants, for example, to provide more information to the people who pull orders in the warehouse. She is hoping that software can be written to provide the warehouse location of goods being collected for an order. Currently, this location information is largely kept in each person's head. If location information could be listed alongside traditional information (e.g., what the item is, how many are needed, whose order it is), time could be saved in training new warehouse personnel. New employees do not have to learn where everything is located in the warehouse. Carla also wants to provide additional information to her customers about where their goods are at any given moment (e.g., in the warehouse, on the trucks). She also sees the stevedores themselves as part of the information network and is thinking of ways to better provide information to them. Instead of having the stevedores come to the docks to find out if there is work, for example, maybe she can use automated phone messages or automated dialing to notify stevedores about the work situation. Carla nows sees workplace automation as having a role in almost all parts of her organization, not just to generate payroll and warehousing paperwork.

Both Carla and Chuck used the occasion of automation to rethink business processes. A workplace computing technology transfer project can, therefore, lead to business process reengineering [8.6–8.8]. But such a technology transfer process almost always retools the people involved in the project. And by retooling the people, the process retools the organization itself. The retooling of individuals means that they have acquired new skills and new ways of thinking. Those involved in a workplace computing technology transfer process develop (or increase) computer literacy. They also gain a better understanding of what others in the organization do and what the organization itself does. They view the organization in new ways. They view the information within the organization in new ways. Because of such retooling and because both Chuck and Carla's projects met final expectations, both projects are a success. Both companies are still on their feet and neither has as much anxiety about surviving as they once did. Both projects are judged to have been worth the time and effort that went into them.

Conclusion

We wanted to end this part of the book with something flashy, but the truth is that technology transfer projects just end. No flash. You can certainly create a flashy event (e.g., a party) if you want, but in most cases the moment after a project ends looks a lot like the moment before. Perhaps this is the most important point about technology transfer. It is a process, not an event. The sum total effect of technology transfer can be dramatic, but any individual moment is probably not. Technology transfer is more like water than lava. The long-term effect of water eroding rock can be dramatic (e.g., the Grand Canyon), but any particular moment of erosion is not. The effect of hot lava is very immediate and very dramatic. So are volcanic explosions and large earthquakes. The uneventful nature of technology transfer is important because we have known too many people who seem to find "eventless" processes meaningless. Such people cannot see progress without events. If there are no events, then — as far as they are concerned — nothing is going on. Plodding along, getting the job done without a lot of events is tough work. So maybe the flashiest way we can end this part of the book is with a short tribute to those who change the world by getting the job done without much fanfare and events. To the World Health Organization, which eradicated smallpox; to those who strung the telephone wires; and to all those other "drips of water" past and present who made an assortment of technologies possible, common, and everyday — usually without accolades, congratulations, and testimonials: thank you good and faithful friends.

References

8.1: Magrass, Yale, and Upchurch, Richard L., "Computer Literacy: People Adapted for Technology," *Computers and Society,* Association for Computing Machinery (ACM) Special Interest Group on Computers and Society (SIGCAS), Volume 18, Number 2, April 1988, pages 8–15.

8.2: Rosenberg, Richard S., *The Social Impact of Computers,* Academic Press, San Diego, California, 1992.

8.3: Currid, Cheryl, "Corporate Computer Illiteracy Can be Addressed by Simply Training Users," *InfoWorld,* February 3, 1992, page 52.

8.4: Lidtke, Doris K., "Computers in Schools: Past, Present, and How We Can Change the Future," *Communications of the Association of Computing Machinery,* Volume 36, Number 5, May 1993, pages 84–87.

8.5: "What Role for Technology? Is Technology Part of the New Standards?" Special Report: Standards in *Electronic Learning,* March 1993, pages 18–19.

8.6: Davenport, Thomas H., "Process Innovation: Reengineering Work through Information Technology," Ernst & Young Center for Information Technology and Strategy, Harvard Business School Press, Boston, Massachusetts, 1993.

8.7: Schnitt, David L., "Reengineering the Organization Using Information Technology," *Journal of Systems Management,* January 1993, pages 14–20,41–42.

8.8: Avishai, Bernard, "A CEO's Common Sense of CIM: An Interview with J. Tracy O'Rourke," *The Information Infrastructure,* Harvard Business Review Paperback, Number 90078, Harvard Business School Publishing Division, Boston, Massachusetts, 1991, pages 61–70.

Chapter 9
Outside Consultants

As a matter of business it was a good idea to get the notion around that the thing was difficult. Many a small thing has been made large by the right kind of advertising.

Mark Twain in *The Connecticut Yankee in King Arthur's Court,* Chapter XXII, The Holy Fountain.

"Language of Chairs, Hyde Park, London, England," photo © copyright Christi Carter, 1993.

In this chapter we discuss computer consultants: whether you really need them and, if so, where to find them. We also list purchase guidelines that may help you in your purchase of computer products. An important thing to remember about some consultants and some computer salespeople is that some make their money by promoting the notion that workplace automation or purchasing computer products is difficult. Some make their money by making a "small thing ... large by the right kind of advertising." They make the ordinary seem unusual or extraordinary. Another thing to remember is that people do or do not use consultants for a variety of reasons. A major reason people do use consultants is to have someone to blame or fire if something goes wrong. ("It is not my fault, the consultant was supposed to know this issue.") Of course, even under the best of circumstances, consultants are hired to be fired. A company should also remember that consultants do not arrive knowing everything. Even the best automation consultant has to spend time learning about the company and how it operates. Sometimes it seems to take a consultant longer to learn about the company than it would have taken someone in the company (who knows the company well) to learn about computers. The organizational information and the automation information need to be blended to create a smooth automation-organization literacy and a technocentrism based in organizational need.

When using consultants (of any kind), it is also best to remember that old mining saying: "Some came to mine gold. Some to mine silver. And some to mine miners." Those who mined miners were those unscrupulous people who made their living by selling equipment and supplies — often unnecessary and of inferior quality or both — to miners at grossly inflated prices. In the computer industry as in other industries, the quality of the product or service is not always directly related to the cost. Quality does cost, but the most expensive is not always the best. So how does a company keep from getting mined? By following the outside consultant guidelines discussed next.

Outside Consultant Guidelines

A set of outside consultant guidelines are listed here that help a company decide when and where to use a consultant, if at all. Following the guidelines should also prevent a company from becoming part of some consultant's mining operation.

Consultant Guideline 1: Automation projects do not always require the use of an outside consultant.

A company can complete much or all of an automation project itself without the use of an outside consultant. The extent to which a company uses outside, paid help and the type of paid help depends on both the amount of expertise within a company and the amount of time the company can spend identifying, collecting, evaluating, and integrating the necessary additional information and expertise. A company may do the more straightforward tasks itself and hire a consultant to do the more difficult tasks. Some organizations hire consultants because the skills and information required are short term and infrequent. The company may not see the permanent support and development of such skills within the organization as cost effective. An example might be use of an outside marketing consultant by an organization that only occasionally undertakes a marketing campaign. A consultant could be the facilitator for the project.

A company may also do most of the work itself and use a consultant occasionally to ensure it is on the right track. Some people we know, for example, enjoy working on their house, but periodically pay an architect for one hour of professional consultation just to ask and receive answers to specific questions. The homeowners do not use the consultant to help them frame their questions and the relevant issues. The homeowners ask the architect only to answer specific questions and to provide specific information missing in their plan for their house. What they are paying for is the vast amount of information available in one place (in the architect) and believe it would cost them at least as much in their own time to develop the answer they seek.

The automation equivalent might be for a company to define its automation needs and then ask a consultant for a range of computers that might do the job. A company might also use a consultant to answer specific questions about the operation of a piece of software. An automation consultant could also help a company streamline procedures and organize data in preparation for automation. So a company can choose not to use a consultant, choose to use a consultant for only parts of the project, or use a consultant for the entire project. Remember, however, that the consultants work for the company. The company does not work for the consultant. Consultant decisions and choices should be reviewed by relevant personnel. Some member of the organization should have the job of managing the consultant.

Consultant Guideline 2: If the company decides to use an outside consultant, choose one the same way any other professional consultant is chosen. Automation consultant selection issues are the same as selection issues for consultants in any other area.

The purpose of this guideline is to remind a company that there is nothing mystical about hiring a computer consultant. A company should use the same decision strategies and approach when hiring a computer consultant as are used when when hiring any other kind of consultant (e.g., architect, attorney, management consultant, marketing consultant). There is nothing unusual about the process of finding a computer consultant. How can a company find a computer consultant? In a variety of ways. Ask friends and professional associates. Ask each consultant you find who is the best consultant for the job other than themselves. Do not ask them who is the best in the field or who is the person most suited for the job because they will most likely respond that it is themselves. Ask them whom they would hire if they needed a consultant and why. There are many things to consider when choosing any consultant.

Among these considerations are the consultant's approach (Does the consultant conceptualize issues in a way that is productive for the company?), expertise (Does the consultant seem to know how to implement the chosen approach?), the consultant's ability to work well with company personnel (Does the consultant have personalities and attitudes that would make the collaboration productive?), the ability to meet any deadlines (Do you have some evidence to indicate that the consultant can complete the assigned tasks within a reasonable length of time?), and is the consultant truly interested in making some aspect of your life/organization more rewarding/productive. Ask potential consultants how they view your organizational problem. Ask to see examples of their work and to speak with other customers. This does not mean that a company should never choose a consultant who is new at the game. The newest consultant could be the best. A consultant in the business many years may not be very good. Remember when choosing and working with a consultant to use your own good judgment. If something does not make sense to you, then ask questions. If the consultant cannot explain the answer adequately, then consider someone else. If they can help you solve a problem, then it should be straightforward for them to explain the problem and solution to your satisfaction.

Consultant Guideline 3: Do not relinquish decision responsibility and control to the consultant.

The consultant's job is to identify the issues and make professional suggestions. All final decisions should be made by the relevant members of the organization and not the consultant. Take responsibility for all specifications; tell the consultants what the company wants. Make sure that you ask for what you want (a familiar computer adage is that people often get exactly what they asked for, but not what they want). It may be tempting to pick a consultant who prefers to make and implement most decisions alone without talking to relevant people, but that path usually leads to unhappiness, the same kind of unhappiness that you might feel if you left all decisions to an architect. You would not just ask an architect to design a house, build it, and then let you know the date you are to move in. When working with any consultant, the company personnel should also get as much information as possible on their own and not just rely on everything that the consultant tells them. A consulting relationship is a partnership.

Consultant Guideline 4: If the company uses an outside consultant for any part of the automation project, establish the total project price, project specifications, completion dates, and deadline details first!

Deadlines and prices should always be specified in a company's contract with any automation consultant. If such details are not specified, misunderstandings and hard feelings could result. The company might also lose control over the course and cost of the project. The company and the consultant could go through life together. The consultant may be continually involved in the project. Consultants may continually be involved because they want to be involved in every automation-related decision. A consultant may also be continually initiating major changes to the system. Having outside automation consultants engage in the first form of continual involvement is usually unnecessary. It is the functional equivalent of having either the architect or the builder give a professional opinion at professional prices on all aspects of the completed house including where to hang a painting and the lamp to choose.

The outside consultant who continually initiates changes is functionally similar to the architect or builder who continues to make changes to the house even after it is complete. You do not expect to move into a house and then have the architect visit every day with new plans for an adjustment to the house — adding a room, eliminating a room, changing the basement. You may plan to make major changes to the house in the future, but the occurrence of the major changes is not considered part of the original contract. You expect a livable and suitable house even without major changes. You expect to be self-sufficient in the use of the house as originally designed.

Specify a date by which the system will be "operational" (working as intended). Specify what you mean by operational. Both the company and the consultant need to know their respective responsibilities in making the system operational. The company should also clearly understand what happens to any products developed as part of the project. Who owns any customized software or other products developed to meet company needs? The company? The consultant? Are they owned jointly? A company wants to ensure that it does not pay the development costs for a product the consultant subsequently owns and sells.

Consultant Guideline 5: Avoid packages of odd hardware and software that only the consultant can operate/repair/distribute. The automation project should not produce future revenues for the consultant.

No company wants the consultant to be the only source of parts (no second source). Do not let the consultant buy equipment for the company. Often the consultant who buys equipment registers the equipment under the consultant's name. The warranty is, therefore, not in your company's name. Upgrade information is sent to the consultant and not your company. If there is a problem, your company may have to pay consultant time to gain access to the manufacturer. The consultant has, therefore, constructed the system in such a way that all adjustments no matter how small genuinely require further involvement of the consultant. Having the outside consultant continually involved and having all the decision responsibility prevents the organization from becoming self-sufficient. If the company purchases the equipment directly, then it will get the information free, it will not have to pay the consultant. The other problem when the consultant purchases the equipment is that your company may pay too much. Consultants usually charge the list price for equipment. The consultant may have gotten a discount that your company could have gotten itself by buying direct.

Consultant Guideline 6: Avoid outside consultants who also sell hardware and software package deals. Also avoid high-cost "training" packages.

Lumping hardware and software costs together in packages is often used to hide unnecessary customer fees. The best buys are almost always sold separately (there are exceptions). Some people who call themselves automation consultants are really salespeople selling one particular configuration (package) of hardware and software. They come in and generally charge consulting prices and at the end of the process they find that their system is the best one for the organization. You can get sales advice free from a vendor of the machine or software so why pay consultant prices for it? You want to avoid consultant package deals for the same reason you want to avoid letting the consultant buy any automation equipment. Letting the consultant buy any equipment, especially a package, can be more costly and make you more dependent than if you bought it directly. Another way some consultants make money is with training packages. They teach members of your organization how to use specific software or hardware. If you do get involved in a training program with a consultant have the consultant outline exactly what is to be learned in the program. What are the educational goals of the training? What guarantee do you have that the material is being taught in a way that your company employees can learn it? Also decide what is to happen if at the end of the training company personnel associates have not yet learned the material. Does the consultant give the training again at no cost or is it offered again at cost? Is the consultant providing training or a course? Consultant relationships should be relatively short lived.

Some consultants make money from training packages you either can get cheaper elsewhere or do yourself. If you want to know how to use a widely used software package, check with your local adult school or university computing center before hiring a consultant. These places often offer computer training at reasonable or inexpensive prices. Training is important, but make sure that you get good value for your training dollars. Some consultant training courses we have seen are based almost entirely on the tutorial that comes with purchased software. The consultant provides structure and a place to work, but no new material. Other training classes do not offer any hands-on experience. The class is lecture. Since the best way to learn to use hardware or software is to work with it, lecture courses have limited benefit. Some classes provide general introductory information. Often you need answers to specific questions and these are not usually addressed in a general course. Make sure you know precisely what the content and method of the course is before enrolling. Make sure you know what you want to buy (e.g., structure, answers to specific questions).

Consultant Guideline 7: Avoid outside consultants with special or narrow organizational interests and expertise. Choose consultants who consider the organization as a whole and not just one aspect.

Consultants and individuals with narrow interests are likely to sell your company what they know and not what you need. This is as true for automation as it is for any other area of expertise. If you hire a French pastry chef to create desserts for your party, for example, you will most likely get French desserts rather than Austrian. Why? Because this is the expertise the paid professional has. Do not assume that because a consultant has expertise in one automation area (e.g., accounting software) that they know all automation and all organizational functions. It is unrealistic to believe that your accountant or banker is also a brain surgeon. So do not automatically believe that your company's outside accounting firm, for example, knows nonfinancial automation, all organizational systems, or how the systems work together. The firm may know this information. It may not. Before hiring any consultant carefully explore their range of expertise.

Purchase Guidelines

Listed next are some purchase guidelines to help make purchasing automation products a little easier.

Purchase Guideline 1: Choose the product of greatest utility. (For computers, this means choosing the software before choosing the hardware.)

This guideline seems almost silly, but companies often do not use good consumer practices when purchasing computing equipment. If this guideline is not followed, then you may purchase a product or service that is not productive or useful for you. To implement this guideline successfully, you must know both what is best for you in detail and the relationship between the product under consideration and the desired results. Individuals are sometimes unhappy with their purchases because they believed that the purchased item had characteristics that it does not, in reality, have. Sometimes the buyer did not fully understand the capability of the product. The buyer failed to evaluate the characteristics of the product fully. In other cases buyers do not fully evaluate what they intend to do with the product. The individuals have not fully defined either what they intend to achieve or what product characteristics are required to achieve the defined goal. The buyers did not fully evaluate the characteristics and functions that are best for them. You can avoid such problems by following the proposed technology transfer steps. You can also avoid such problems by choosing your software before you choose your hardware. Let your software needs determine the range of hardware choices. Do not let your hardware choice determine your range of software choices. The software is what does the work for you. Once you buy the hardware you are tied to the software that runs on the chosen machine and the software you want may not run on your chosen machine.

Purchase Guideline 2: Try to purchase a computer with the greatest compatibility. Pay careful attention to the computer's processor.

Purchasing a computer with the greatest compatibility means picking a computer (regardless of brand name) with the largest market share. Compatibility is primarily determined by the microprocessor. Machines with the same or a similar microprocessor are usually compatible. Software specifically written for one machine will run on a compatible machine. Compatibility is important because software, second sources (sellers), and parts are more likely to be available for the machine with the greatest market share. Second sources are important because they offer you a place to take your business if you become dissatisfied with either the pricing or service of your initial source. The more software that is available, the more likely it is that as your needs change the software you need will be commercially available. As you begin to use a computer you will most likely find new uses for it. These additional uses may only become evident after working with the computer for a while. The new uses may require new software. The better the computer's compatibility, the more likely it is that the desired software already exists in commercial form. Remember that the word "compatible" is sometimes used to describe less than total compatibility. That is, machines are often termed compatible when only some of the software specifically written for one machine will accurately run on a compatible machine. So if someone says that the machine is compatible with another, ask how compatible.

Purchase Guideline 3: Include both short-term and long-term costs in your calculation of total product cost.

Your total cost is not just the stated purchase price of a product, but the total cost of acquiring and using the product or service productively in your life. This is not a new concept. You should calculate, and most likely do calculate, such costs for all products that you buy. If you do not, you probably spend more money than you originally anticipated. In choosing an air conditioner or a refrigerator, for example, you probably look not only at the purchase price of two comparably equipped models, but also the efficiency rating and maintenance costs. The higher the efficiency rating, the less electricity used by the product to cool a room or food. The smaller the amount of electricity used, the cheaper the product is to operate. Thus the machine with the higher efficiency rating is cheaper to own and use. It may even be that a machine with the more expensive purchase price has such a superior efficiency rating that when the price is calculated over the first year of use it is cheaper to purchase and use. With automation products, you need to consider not only the purchase price, but also the long-term costs of operation, maintenance, and upgrades.

Purchase Guideline 4: Look for the hidden costs (e.g., inexpensive hardware and expensive software, inexpensive software with expensive hardware, high maintenance cost).

There are hidden costs in the purchase of automation products just as there are hidden costs in the purchase of other products such as a house. Hidden costs can be both short and long term. The true price of a house to you, for example, is not just the purchase price. The true price includes the sometimes hidden costs of the mortgage (e.g., interest rates, points), insurance, attorney's fees, title searches, taxes, repairs, electricity costs, water costs, and heating costs — just to name a few. You must also consider the cost of using the house in your lifestyle. A house that is not close to where you work can be expensive because of high commuting costs. In the purchase of a computer as in the purchase of any other product there are often cost items that are not expressly stated, but that are necessary to take possession or use the product.

It is important, for example, to consider the cost of software for any machine you are considering. An inexpensive machine with expensive software can be an expensive system to use. An expensive machine with inexpensive software can be an expensive system. A hidden hardware cost may be the cost of altering your company environment (e.g., adding more air conditioning). You need to know whether a machine under consideration requires a special environment. Is the current environment in your office conducive to the operation of your chosen equipment and, if not, how much will it cost to adjust the environment? You need to know the cost of computer options/parts and who is required to replace or install options/parts. Can you replace parts yourself or can only a professional service person make hardware adjustments? How soon will you have to replace the machine? How often does the machine break (mean time between failures)? How long will its design and equipment be current? How long do you keep a machine? Companies sometimes keep a machine as long as it does the job. Just because a newer machine is sold does not mean you have to dump your current machine and buy a new one. After all, you do not trade in your car or typewriter just because it is not the newest model. How expensive are supplies? Is there a second source on all parts and equipment?

Purchase Guideline 5: Consider the resale price of purchased software and hardware.

The organization may want to sell the purchased equipment in the future and purchase new equipment. The potential resale value of a product can be considered a hidden cost. If a product has a poor resale value, then this is a cost. The resale value of a product is a common purchase consideration especially if you plan to sell or trade the item at some point. Selling or trading the item is common with large ticket items such as a car or a house. If two products have a similar total calculated price and one is known to have a much better resale value, then the one with the higher resale value may be the better buy. Many things affect the resale value of any item. One is the basic quality and workmanship. A well-constructed item will most likely be worth more than a sloppily built one. A product built with quality material is worth more. A well-maintained product is worth more than one that is poorly maintained. Take care of your automation equipment so that you will be able to get the best possible value for your equipment if you choose to sell. Another consideration is whether the company manufacturing the automation product will still be in business if you decide to sell the product. The products of a company no longer in business will most likely not be valued as highly as the products of a company still in business. Given the volatile nature of the computer industry, the survival probability of a computer company may be an important purchase consideration.

Purchase Guideline 6: Purchase only reputable, high-volume products. Such products are usually of the best quality at the lowest cost.

High-volume products are good for many reasons. The high-volume product is the one most likely to be sold at a good price. High volume usually means there has been adequate use of the product to determine and eliminate certain problems/bugs in the production and operation of the product. The larger the number of product users, the more likely it is that both second sources and a number of service vendors will be available. The high-volume product is the one most likely to have generally available support. Reputable means that the company has a reputation for building good quality products with solid construction. This guideline does not mean that you should never take a chance on a new product. It only means that you are taking less of a chance with a high-volume product from a reputable company. When you buy a new product, especially one from a new company, you must guess the reliability of the company, the reliability of the product, and the eventual popularity of the product. When you purchase a reputable, high-volume product, no guess is usually necessary. Real information and data are usually available through reviews, consumer groups, etc.

Purchase Guideline 7: Purchase a fairly new product.

Shop around for products and purchase a fairly new product — one with fairly new technology. Fairly new technology will be cheaper than the newest technology. New technology is defined as technology that is not obsolete, you can still get parts, software, etc. Do not buy old technology. Old technology may be cheapest, but it is no bargain. Do not buy any Model T's of the computer world. The idea of purchasing a fairly new, but not the newest, technology applies to many products. Some people will not buy a particular model of a car during the first year it is sold. Such people often believe that it takes at least a year of operation for all the bugs and problems to be eliminated. By waiting a year such people can see if the product becomes a reputable, high volume-product. Some people wait for the second year of a product because they expect the price to drop as the product becomes slightly older than the current state of the art. But you do not want to wait many years before purchasing a product. You do not want to purchase an old product. An old car, especially one that has not been maintained, is not a good value. An old house that has not been maintained and renovated to keep pace with current lifestyle demands is not a good value. Buy fairly new automation products with a reasonable life expectancy. Trying to do this can create a tension between getting a well-tested technology and buying a technology with the longest possible life. Clairvoyance is a definite asset when trying to follow this guideline. It would help you buy the technology when it first appeared to maximize longevity (cost is another matter).

Purchase Guideline 8: Consider the full range of product alternatives — even ones that seem extreme — because this will clarify your automation picture for you. Clarification may reveal a low-cost solution that is as effective as a higher priced solution.

The term "full range" applies both to the functionality of the product and the price. Sometimes the higher the price, the greater the functionality, but this is not always the case. The most expensive is not always the best. Look at products slightly more expensive than you want to spend. Remember that because of hidden costs the product that initially looks more expensive may not be so. Sometimes what looks like a huge price differential really is not. Look at products with slightly less functionality and slightly more functionality than you think you need. Consider buying a machine with more functionality or "power" than you currently need if you think your automation needs will grow. When purchasing a computer, consider the range of situations and machines. If, for example, more than one user is going to use the system, consider both a multiple-user system and a distributed network. Examine both new and used products even if you think you want a new product. Examining a variety of products helps you to clarify your thinking. You may also find a low-cost alternative that works as well as a higher priced alternative. As you examine each product, consider it in light of your requirements and consider your requirements in light of the characteristics of the product. Have feedback loops in your product consideration process. There is nothing wrong with examining a product that you initially consider unsuitable just to get a sense of the range. By looking around you become better able to recognize value.

Purchase Guideline 9: Purchase one type of machine.

When automating with more than one computer, try to purchase only one type of machine and one type of software. Strive for uniformity. This should decrease the cost of repairs and training. Uniformity also means that you almost always have a working machine through $n + 1$ redundancy. When you automate using more than one machine (e.g., in a distributed network), try to use the same machine throughout. Uniformity is as valid a concept for workplace automation as it is for any other product. You do not want to use different makes and models of automation products within one workplace automation system. Try to purchase all items of the same kind from one manufacturer. Doing this has many advantages not the least of which is the decreased cost of training. There is only one model to learn and people within the organization can train others within the organization. Another advantage is that you almost always have a working machine because you can replace a broken machine with another working machine. Using one product throughout may also allow you to purchase in bulk.

Purchase Guideline 10: Do not buy more than you need.

Do not buy more services from a vendor (store) than you need. Computers, like other products, can be bought at discount prices from stores that do not provide the service or environment of higher priced stores. Although there are many computer vendors, vendors are usually of one of two types. Discount vendors just sell the product. The environment of a computer discount store may not be glamorous. The full-price vendor usually sells a large range of services in addition to the product. The environment in a full-price store is usually better (e.g., better furniture) than that in a discount store. Which type of vendor you choose depends on what you want to buy. If you know exactly what you want and are interested in the product alone, then a purchase from a discount vendor is an excellent idea. You may also wish to purchase from a discount vendor if a full-price vendor offers services that are not important to you. The discount vendors are usually cheaper because they do not have the same overhead (e.g., lower quality environment, no extra personnel for the additional services). In general, discount computer houses do not sell service. By service we primarily mean the service of being able to discuss your computer needs at length with personnel and to call them with questions if you run into trouble when using the equipment. This does not mean, however, that all discount vendors refuse to answer questions. You should also know that many computer products come with a hotline telephone number. Such a hotline is operated by the manufacturer. Sometimes the hotline telephone number is toll free. So even if you purchase from a discount vendor and are not using a consultant, there may be someone to call to get answers to your questions.

Some of the full-price vendors, however, will also let you test a piece of software if you have done previous business with them. You can go to the store, test, and evaluate a piece of software without having to buy the software. This is convenient if you are considering a new piece of software and do not know anyone who owns the software. If you buy from a full-price vendor, however, make sure that the vendor really can and will provide the promised support and service. We have met some full-price vendors who are arrogant and unable to answer many basic questions. If you are going to buy advice along with the equipment, ensure that you like the people, that you can work with them, and that they seem to know what they are doing.

Purchase Guideline 11: Try to buy in quantity or bulk since this usually reduces the cost of any individual item.

Buying in quantity is not a new concept and is applicable to a variety of situations from the purchase of chicken legs in a grocery store to the purchase of computers in a computer store. Many computer vendors consider "bulk" or "quantity" to be the purchase of three or more of the same items at the same time. For such bulk/quantity purchases there is usually a discount. Some vendors discount further as you increase the number of units purchased at the same time. Not all vendors give bulk discounts, however, so when you are shopping for vendors ask each potential vendor if they do.

Chapter 10
Computers:
Myths and Realities

Somehow, every time the magic of folderol tried conclusions with the magic of science, the magic of folderol got left.

Mark Twain in *The Connecticut Yankee in King Arthur's Court,* Chapter XXXIX, The Yankee's Fight with the Knights.

Myth, magic, truth, and reality: When hearing about a technology (especially a new technology), it is often hard to tell which is which. It is, however, important to know which is which. When it comes to conclusions, when it comes to a technology's performance, magic and myth usually get left behind truth. If you believe the myth, therefore, you will probably be very disappointed with use of the technology. That is why, in this chapter, we discuss some myths we have heard about computers and try to dispel them. Maybe one or more of the following myths is one you currently believe or once did.

Why do people believe myths? In some cases the myth is the only information a person receives and it seems like plausible information. The myth information does not seem too ridiculous to be rejected — it might be true. In other cases, real information competes with myths. Both are available. When both are available, which one "wins" the competition? — Often the one seen as more affordable (in terms of time, money, and effort). All information/beliefs have a price associated with them. Time, money, and effort are needed to acquire and use the information. Once the sixth-century residents believed the Yankee's technologies to be magic, for example, there was little left for them to do. Magic does not require additional investigation. Magic does not challenge human understanding. And sometimes people believe myths because, well, it would just be so wonderful if they were true.

Myth One: You have to be a genius to work with a computer.

This myth is dying, but can still be found. It is dying largely because computers are becoming easier to use and because children are using computers. It is difficult to sustain the idea that computers are difficult to use when three-year-olds use them. It is true, however, that almost everyone who has learned to use a computer is perfectly willing to continue this myth. The scenario runs something like this: "I am not a genius so I cannot learn to use a computer for anything. Hey, almost in spite of myself I am learning to use a computer. It is even fun. Conclusion — I must be a genius after all." One result of computer use seems to be an increase in the number of people calling themselves geniuses. Many of those who believe this myth, however, do not even attempt to use computers. Such people usually say, "I am not a genius. It is hopeless to even try to use a computer."

Another effect of the myth is that some of these geniuses often try to stay geniuses by intimidating others. We have seen geniuses explain the simplest of word processing commands in the most convoluted manner. These genius instructors do not intend for the student to understand the commands and often seem to take some satisfaction in showing the student repeatedly how to accomplish the task (because they can humiliate the student each time). So if someone teaching you about computers seems intimidating, rude, and arrogant, it is not a sign of genius. It is a sign that the person is intimidating, rude, and arrogant.

So if it does not take genius to work with computers, what does it take? Time and effort. How much time and effort? The same amount of time and effort required to learn any new subject: 15 minutes and the rest of your life. In 15 minutes you can learn the basics of computer use (or driving or mathematics or Japanese, etc.). You will, however, continue to learn new things and increase your computer proficiency every time you use computers for the rest of your life. And you must use computers to get better at using them. The way you get to Carnegie Hall and the way you get to be a "genius" with computers is the same: Practice! Practice! Practice!

Myth Two: You have to know higher level mathematics to work with computers.

Like the "genius" myth, this myth is also dying. To use a computer you do not need to know higher level mathematics unless the application (the software you are using) requires higher level mathematics. You do not need an extensive science background either. If you believe this myth and do not know higher level mathematics, you may never learn to use a computer. You would have to learn all this mathematics and science first. What do you have to learn to use computers? You will have to learn to be precise and logical if you do not already possess these traits. The myth of needing higher level mathematics seems to have been born in the days when computing meant building the machine itself. Designing and building a machine does take a mathematics and science background. This myth may also have gained credibility because most early machines were used to do higher level mathematics and because most university computer science departments were originally located in either the mathematics or electrical engineering departments. But if you are not going to design and build machines or use them to do higher level mathematics, you do not need to know higher level mathematics.

Myth Three: Computers are like nothing ever seen before in the history of the world.

In many ways computers and computing are unique. In other ways computers are machines just like any other machine. In another sense computers are not "new" because they only undertake functions that are well known — functions that are not new. Computers are primarily extensions of current activities. Computers are best at performing tasks that are well known and well understood. Why? Because computers only follow detailed instructions. These instructions are given through the software. Because of the level of detail required by a machine it is not really possible to have a machine do anything that is not well understood by a human. This does not necessarily mean that you must understand the phenomenon, but someone must. Well, if the computer functions are not new, then perhaps the effect of computers is new? Not really. The effect of computers is no more startling than the effect of any other technology.

Every new technology is strange, amazing, and familiar all at the same time. Is traveling across the country by airplane different from traveling across the country by either car or horse? Yes, in some ways. Airplane travel is in some ways more comfortable, less aggravating, and less boring than either automobile or horse travel, but airplane travel can be uncomfortable, aggravating, and boring. It was once exciting to fly. Now flying is no more exciting than riding a bus. Computers used to be more exciting too. You used to be able to entertain people for hours with stories about your computer. Now, in most cases, people would rather hear about your flight.

Computers are also similar to airplanes in that each technology allows you to spend more time on the "interesting" tasks. Each technology in its own way allows you to do things that were not possible without use of the technology. Computers allow you to spend more time on the work itself and less time on the routine tasks that must be done to deal with the work material itself. Word processing software, for example, allows you to spend more time with document content because you do not have to spend as much time with document format (e.g., size of margins, position of headings) and typing (and retyping and retyping and .....). When people traveled across country by horse, much time and energy was wasted in the sense that the activity of traveling across the country was not usually the goal. People were traveling because they wanted to get someplace. Airplane travel allows you to spend more time on the desired activity (vacationing, business) and less time getting there. The lifestyle changes resulting from airplane and computer technologies have made life different. In some ways that difference makes life better. In other ways it does not.

Myth Four: Using computers means that everything will be done by the touch of a button.

Even today, some people really believe this myth. Belief in this myth is often revealed by such statements as "This information is on the computer. Someone should be able to get it to me in five minutes." It is accompanied often by a deep annoyance when people say it will take time to generate the report or analyze the computer-based information. You also see belief in this myth revealed by the belief that people working with computers do not need to be paid very much or be highly skilled because all they do is "push buttons." But little, if anything, is done by the touch of a button. Those who believe this myth greatly underestimate the amount of time, effort, and button pushing required to work efficiently and effectively with computers.

Myth Five: Computers never make mistakes.

Computers commonly have the wrong information or produce the wrong information, but a human not the computer usually made the mistake. On occasion a computer may suffer hardware damage that causes computer-generated errors, however, almost all so-called "computer errors" result from human error. There are two main sources of these human mistakes/errors: the end users and the software developers. End users (people who use the computer on a daily basis) usually create errors either when entering data or when entering commands necessary to use a software applications package (e.g., word processing, spreadsheet, statistical package). A data entry error occurs when a person enters the wrong information (e.g., wrong last name, misspelled last name, wrong amount) into the machine. A command entry error occurs when a person enters the wrong command to accomplish the intended task with an applications package. When an end user sees a computer error, there is a tendency for the end user to immediately say "the computer is broken." Usually the computer is not broken and the end user can fix the problem (e.g., reenter data correctly, reenter a command correctly). Sometimes, however, the computer is "broken" because the error results from a programming error.

A programming error occurs when the person writing the application gives the wrong instruction (writes the wrong line of code) to the machine through the software. Such an incorrect instruction may tell it to use the wrong piece of data or to execute the wrong calculation for the task. Programming errors (also known as "bugs") can be caused by carelessness. They can also be caused by the sheer size and complexity of a program. Sections of a program may interact "unexpectedly" or in unexpected ways with other sections of a complex program. The larger and less used a program, the more likely it is that there will be bugs. No programmer who has worked on a large project who on hearing that something is not working well does not freeze in his or her tracks or wake up screaming in the night wondering if it was a misplaced comma or some other problem. It is expensive to debug code properly and many places rely on the users to do it for them. Debugging is like running a new chemical product through a series of tests before approving it to see what the side effects are. There is no single process for code that does this.

Today's computers rarely make mistakes with the information they are given, but they do produce information errors. If given the wrong information, computers will use and report wrong information. Computers also produce errors a human would not because computers work only with the information provided and the information provided to a computer is limited. Humans have more information available for use than is provided to a computer and, therefore, can check the accuracy of a piece of information against more information than can a computer. An example of a common computer error that would not be produced by a human is the salutation in a computer-generated letter "Dear World Trade Center." A human would not generate such a salutation because most humans know that the World Trade Center is not a person and that only human beings are addressed in the salutation of a letter. All the computer "knows" is that it is supposed to place information from a particular computer location onto a specific location on a piece of paper. The salutation error was not caused by the computer. The error was probably caused by a human who placed the term "World Trade Center" into a computer location meant to contain only the names of people.

Myth Six: Computers think.

It should be clear by now, however, that computers do not think. Human beings think, but computers do not. A computer is a model — and currently not a sophisticated model — of the human mind. Unlike a human, however, a computer does not "invent" things. There are efforts to make computers think, but the thinking computer will not be possible for a long time — if ever. Why not? Because human thinking and creativity are not understood well enough (and probably will not be for a long, long time) to give computers the detailed instructions necessary to make them think and be creative. Computers are better used to assist humans in thinking and being creative.

Myth Seven: Everything computerized is great.

There is nothing about computers in and of themselves that makes computerized tasks great or guarantees increased information efficiency and effectiveness. Increased efficiency and effectiveness rest firmly on careful choice and adequate, appropriate use of good computer products. When good computer products are used properly, most things computerized are great. If not, things are miserable.

Myth Eight: The automated office is a paperless office.

The automated office often seems to have more paper than the nonautomated one. Why? Because with automation comes reports. Many reports. And these reports come on paper. There are reports used to check data accuracy. There are reports that you could not have produced in a nonautomated environment. These reports usually analyze or combine data in ways not done under a manual system. The paper that is usually eliminated is the paper used to record and store data in a nonautomated environment. The automation is only worthwhile if you do something with the data stored in the computer. "Doing something" with the data means reports and this means paper.

Myth Nine: Computers are isolating.

We have never found computers to be isolating. Rather than being isolating, we find that working with computers often makes you yearn for isolation. If anything, you meet more people when you use computers. Using computers always means solving problems (especially for new users). These problems are almost always caused by some detail you overlooked and often seem to defy solution. After giving a problem your best shot, you usually find yourself desperately calling someone to see if they have ever run into the same problem you have and, if so, how they solved it. Solving a problem can be so frustrating, it makes you bold. You call or stop people you have never met because you heard they solved the problem you are experiencing. If you write software, you often find yourself hunting down friends, strangers, and sometimes even enemies alike to show them what your software can do. And if you write software that others use, you are frequently contacted by people who need information about your code (software) or have found what appears to be a bug (people love to tell you about "your" bugs).

Using workplace automation often means coordinating your activities and requirements with others to a greater degree than in a manual work environment. This increased coordination usually means increased contact and discussion. You will also find that if you use computers correctly you probably increase the number of people you know and with whom you have regular contact. One group of people you want to establish contact with are people with similar automation interests and needs. People with similar automation interests are those using or planning to use the same hardware or software. Those with similar needs are those who have automated or who would like to automate the same organizational tasks. Establishing relationships with others who are automating or who have automated can help everyone be self-sufficient and independent of high-priced "experts." People with the same interests and needs are usually very open and friendly about sharing their information and expertise (within reason, of course, and if you make it clear that you will share helpful information with them and others). Others in your industry using automation are the best sources of adaptation information, which is information about adapting the automation to daily functioning and vice versa.

Makers of automation equipment may know their equipment, but they are not usually the best source of adaptation information. Although much of the communication between those of similar automation interests is informal, sometimes it has a formal structure such as a users group. A users group usually contains people using the same software or hardware in the same industry although sometimes members of the users group are from many industries. Some users groups provide formal input into a computing company on enhancements (improvements) users would like to see. If your automation has a users group, join it (you do not have to go to all the meetings). If not, think about starting one (and holding meetings in places you always wanted to visit). In any event, make friends and trade information.

Some aspects of computer use — such as use of electronic mail — actually open up additional modes of communication. Electronic mail (E-mail) is mail sent by one person via computer to another person via their computer. E-mail has many advantages over telephone or paper mail. It is more immediate than paper mail. A few seconds after you finish typing and sending your message, it can be read by the person to whom it was sent. People also tend to send information and communications over electronic mail that they would not send via telephone or paper mail. E-mail may not replace telephone or paper mail, but it may increase communication. You often find yourself writing to people you would never call or send a letter. You also get messages from people who never called or wrote you before. E-mail is also effective for sending immediate messages that you do not want to communicate in person (e.g., over the telephone). If, for example, you are not going to attend the 11:00 a.m. meeting (and it is now 10:30 a.m.), and you do not want to get involved in a telephone conversation with the meeting chairperson explaining why, you can just send the chairperson E-mail and not answer your telephone for the next half hour.

E-mail is also effective in contacting people who are hard to contact. E-mail communication is even effective with computer hackers who spend most of their time in windowless offices in front of a computer. You send the message. The person's computer gets the message and the person can read the messages at his or her convenience. Using E-mail is also an effective way to control message interruptions.

If many people contact you during a work day and you need quiet time to work, you can turn off your telephone and tell people to E-mail their messages. When you are ready to take a break from your work, you can read and respond to their messages. E-mail is a major component of the information superhighway (cyberspace) [10.1–10.5]. See Figure 10.1 for some additional types of software and examples of when you would want to use them.

Reasonable Computer Expectations

So what can you believe about computers? What should you expect from computers? What you can believe is that computers do provide some distinct advantages:

- Expect computers to have greater information capacity than manual systems.

- Expect computers to handle information faster than manual systems.

- Expect computers to be more accurate and precise than manual systems.

"Greater information capacity" means that you can reasonably store more information with an automation system than with a manual system. Compared to paper storage, the same amount of information can be stored in a smaller more protected environment when using computers. Information contained in an entire room full of file cabinets can, for example, often be stored on one computer disk. "So what??" you ask, "How is this an advantage?" Well, one advantage is often reduced record storage costs. The reduced storage area for automated information reduces storage costs — less space is required. Because a greater amount of information can be stored in a smaller area it is also possible to store greater amounts of information than before. If, for example, you really only have room to store a 5-year history of paper you may be able to store a 10- or 20-year history on disk if you so desire. The ability to store more information easily can also make the information secure from accidental loss. A copy of important automated information can be easily made and kept in a place separate from the original. Try doing this with thousands of pages worth of paper material.

Figure 10.1 Software Products: Types, Uses, and Examples

Type: Word Processing
Typical Uses: Memos,
  reports, letters, labels
Examples: WordPerfect,
  Word

Type: Spreadsheets
Typical Uses: Calculations,
  spreadsheets, graphs
Examples: Lotus 1-2-3,
  Excel, Quattro Pro

Type: Statistical
Typical Uses: Statistical
  analysis, charts, graphs
Examples: SPSS, SAS

Type: Graphics
Typical Uses: Images, image-
  editing, illustrations
Examples: Acrobat, Fractal
  Design Painter, Photoshop,
  assorted packages with "draw"
  in their name

Type: Databases
Typical Uses: Organizing
  data, reporting on data
Examples: Dbase, Paradox,
  Approach

Type: Windowing
Typical Uses: Use (go back
  and forth between) many
  applications at once
Examples: Windows

Type: Accounting
Typical Uses: Various
  accounting functions
Examples: Quicken, Profit,
  ACCPAC Plus

Type: Communications/Networking
Typical Uses: Linking machines
Examples: WinFax Pro, Crosstalk,
  Novel, Banyon, DCE

Other Types (examples):

Electronic Forms
Antivirus
Utilities
Flowcharting
Brainstorming
Desktop publishing
Speciality (e.g., medical, legal)

Type: Computer-Aided Design
  (CAD)
Typical Uses: Drafting, design,
  rendering
Examples: MicroStation

Computers are much faster than paper (manual) systems. They perform millions of instructions per second. In fact, computers are rated in terms of millions of instructions per second (mips). Compare mips to the number of "instructions" you give a hand-held calculator in one second (see who can add 100 numbers faster: a computer or you with a hand-held calculator). What does this speed have to do with your workplace life? Well, information stored on a computer can be sorted, manipulated, and reported more efficiently and effectively than can information stored on paper. This feature makes a very large weekly payroll doable. Under a manual system it would be impossible to do all the calculations necessary to complete the payroll, generate the checks, and produce the payroll reports needed each week (generating the needed reports is as time-consuming as producing the checks). The speed combined with greater capacity also makes your information more useful to you. Information is not useful if you cannot get to the information. Paper information is hard to get to because it is so time-consuming to find, record, and analyze required information. Automated information is easier to get to because of the speed with which the computer can find, record, and analyze large amounts of information. Computers can calculate statistical formulas faster than a person. The greater storage capacity coupled with this increased speed has made large statistical studies possible. This speed also makes "what if" analyses possible. A "what if" analysis is one where you visualize what might happen if make a certain decision.

Computers are more accurate and more precise than manual systems. But how can this be possible? Don't they make mistakes? After all don't humans give computers almost all the information they use? Well, yes, but computers have three characteristics that give them the accuracy and precision advantage over most people. One characteristic is that computers are tireless (within reason, of course). Another is that computers are very good at applying the same rule over and over again. The third characteristic is that computers restrict human "spontaneity." How does being tireless help precision/accuracy? Well, computers can work longer hours without fatigue than most humans. They will — within reason — do the same task over and over again the same way. This characteristic eliminates the random fatigue-related errors that humans sometimes produce. This makes computers very effective at performing well-defined repetitive tasks requiring precision and accuracy. A payroll check produced at 8 p.m. is produced with the same vigilance and attentiveness as the one produced 12 hours earlier.

When we say that a computer is good at executing rules, we mean that if you give a computer a general rule (e.g., add 5% to everyone's salary, deduct 2% from everyone's salary until they have a year-to-date earning of $20,000 and then deduct 1%), it will execute that rule consistently. The spreadsheet recalculations just mentioned are a form of rule execution. Anyone who has ever worked with spreadsheet financial information knows how frustrating and time-consuming it is to ensure that changes to one cell are reflected in the changes to other dependent cells. If you are manually working on a large complicated spreadsheet, you often lose track of which cells have been recalculated and which have not. Working with an automated spreadsheet can eliminate this problem. When you start working with an automated spreadsheet, you can specify via a formula (a general rule) the relationships between cells. Then as you make a change in one cell the effect "ripples" through the spreadsheet into all cells using the changed cell value.

What does it mean to restrict human "spontaneity" and how does this help increase precision and accuracy? Often many people use a shared computer system. For them to effectively use it they must agree on the meaning of different fields in the computer system. They must often agree on specific codes to be used for data entry. Any changes they may wish to make may need the agreement of all others using the same information/fields. A major effect of this is that everyone using the system must think about their information more precisely and accurately. Clearly each group of people using common information must have a mechanism for making decisions about changes to the system. (This need for agreement also adds to the nonisolating effect of computers. You are either always talking to people to coordinate your activities, defending yourself against people you should have coordinated with, but did not, or hunting down people who should have coordinated with you, but did not.) Done properly, this can be a productive process. It can force you to evaluate constantly your method of working. It can force everyone in the company to evaluate constantly the company's method of operation.

Another advantage is that quality workplace automation improves not only task effectiveness and efficiency, but also provides a structure for discussing and modifying the flow of organizational information. The flow of information in most organizations is little discussed and addressed and in the absence of automation requirements information flow is difficult to describe. Workplace automation forces an understanding of the information and information flow in an organization to a detailed degree. Such an understanding and the increased or added information that becomes accessible through automation greatly benefit a company. Accessible accurate client, patient, or customer records, for example, make organizational self-examination, self-monitoring, and feedback possible. Such activities allow the organization to have objectivity about its own workings and such feedback can greatly advance organizational knowledge and progress. The medical profession as a whole took a giant leap forward in understanding itself when medical records procedures were developed at the Mayo Clinic. Individual companies can make the same leap when their records are automated.

Another major effect of automation that should not be overlooked is increased personalization. Computers make large-scale personalization possible. Computerized airline reservations systems make it possible to reserve your favorite seat before you get to the airport. Under a manual system such information and the process itself is too much to handle. Automation also increases the "do-ability" of staggered/personal work schedules. The information flow in automated organizations often "looks" the same regardless of whether everyone is working in their office, at home, or in different states.

References

10.1: Ayre, Rick, and Stevenson, Ted, "The E-mail Personae," *PC Magazine,* Volume 13, Number 7, April 12, 1994, pages 251–294.

10.2: Kantrowitz, Barbara, Chideya, Farai, and Biddle, Nina Archer, "The Information Gap," *Newsweek,* March 21, 1994, page 78.

10.3: Meyer, Michael, "The 'On-Line' War Heats Up," *Newsweek,* March 28, 1994, pages 38–39.

10.4: Seymour, Jim, "On-Line Ties that Bind," *PC Magazine,* Volume 13, Number 6, March 29, 1994, pages 99–100.

10.5: Miller, Michael, "The Long and Winding Road," *PC Magazine,* Volume 13, Number 6, March 29, 1994, pages 79–80.

Chapter 11
Computers:
Hardware, Software,
and Vendors

This information did him no damage, because it left him as intelligent as he was before.

> Mark Twain in *The Connecticut Yankee in King Arthur's Court*, Chapter XXVII, The Yankee and the King.

"Statue at the Cathedral at Reims, France," photo from the Cooper Union Library Picture Collection.

In this chapter we provide some basic information about computers and how they work. Some may be familiar with this information. Some not. Some may find it interesting. Some not. But in any event, reading this information should do you no damage. A very basic piece of information is that computers have two major parts: hardware and software. The two parts are interrelated, but different. Computer hardware is the machine itself. Hardware is what you see when you look at a computer. It is what you touch. Computer software, on the other hand, is the part that gives the computer ability, function, and usefulness. Computer software is primarily what you use to complete functions or tasks with a computer. Computer software is the set of detailed instructions programmed into a computer about the execution of a specific task. Hardware is to a computer what the physical body is to a human. Software is to a computer what the mind is to a human.

Although computer hardware and computer software are treated as separate topics, they function together as a unit — the computer system. The performance and quality of hardware and software are intertwined and the general quality of your computer system depends on the quality of both hardware and software. Good software can be restricted in performance by poor hardware. Good hardware can be underutilized by poor software. The term "computer" sometimes refers to the hardware alone. Sometimes the term refers to the combined hardware-software system. This is similar to the two common uses of the word "body" to reference a human being. Sometimes the term "body" refers to the physical body alone. Sometimes it refers to the combined physical-mental system.

Not surprisingly, then, hardware and software developers are different from each other. A good hardware person/company is not necessarily a good software person/company. Why should one expect that they should be? A good psychiatrist is usually not a good orthopedic surgeon. Different skills and interests are involved. There is a symbiotic, but often uneasy, relationship between software people/companies and hardware people/companies. Each group views the other with some disdain believing that the heart of computer science lies in their expertise. Yet a hardware company needs software vendors — large, small, or freelancers — to write good software that runs (operates) on their machine. The greater the amount of good software for a machine, the more likely it is that the machine will sell. A supply of good, available software for a machine makes the machine more popular than one that does not have such a supply.

Software vendors who have written software for a machine hope that the hardware company will sell thousands or millions of machines. The more machines that are sold, the more likely it is that software written for the machine will sell. So it is not surprising that in the area of software availability the "rich" machines get richer and the "poor" machines get poorer. Machines that sell well and have much available software are more likely to have additional software written for them than machines that do not sell well and have little available software. Why? Because quality software is expensive to produce. The production of quality software requires a detailed, in-depth understanding of many subjects, much time, and usually many people. Usually many copies of the developed software need to be sold for the product to break even.

The symbiotic relationship between hardware and software companies in effect makes the market for each other. The software developer wants to see certain hardware succeed and the hardware developer wants to see certain software succeed. Software vendors want to see a particular machine and all compatible machines succeed. A compatible machine is one that can run software developed for another. An IBM compatible computer is one that can run software developed for an IBM machine. An Apple compatible computer is one that can run software developed for an Apple machine. If two machines can run the same software without modification (change) to the software, then they are compatible. Lesser known machines that are compatible with more widely knows machines are often called clones. Compatibility is an important issue since software is not always transportable. Note that compatible machines are not usually exactly alike. Software that runs on one machine does not always straightforwardly run on another machine. Some parts of some software may run well while other parts of the software do not. You should realize that compatibility is not an all or nothing dimension. Compatibility is a very tricky business. To make sure that software is compatible with your machine, you really need to run it on your machine.

The higher the demand for a certain type of software (e.g., financial spreadsheet software), the more likely it is that it will be developed. If not many people or companies need the software, it is unlikely that it will be available commercially. This aspect of the computer industry is similar to the development of orphan drugs in the pharmaceutical industry. Orphan drugs are those with limited sales potential because few people have a disease or disorder that requires the use of the orphan drug in treatment. If not many people have the same disease or disorder you do (and your treatment is not one used by many others), then it is unlikely that your needed drug will be available commercially. It takes as much time and money to develop an orphan drug with limited sales potential as it does to produce a drug with great sales potential. Many pharmaceutical companies develop some orphan drugs, but not all that are needed are under development. If you need an orphan drug, you may not get it. Unlike drugs, if you require "orphan" software, you may be able to develop it yourself or hire someone to write customized software for you.

To understand the nature of customized software, you should know about two other types of software: programming languages and mass-produced applications packages. Customized software is usually known as a customized applications package. Applications packages are what the end user usually sees and uses. Applications packages are written in a programming language, but you do not need to know a programming language to use the package. (As a side note, this statement is not exactly true since some people think that each application package should be a small special-purpose language.)

Programming Languages

A programming language is the way computer professionals communicate with a computer. It is the basic way to give instructions to the computer. A programming language is simply that — a language for talking to computers. New words are often added. The "grammar" may change. New languages are developed. The newer the language, on average, the simpler and more powerful the language. A language becomes simpler and more powerful as the amount of work done by one command from a human increases. Work is done by issuing low-level instructions to the hardware.

At one time the instructions had to be issued one at a time at the same level of detail that the computer works. Instructions had to be entered in the low-level machine language. To issue the instructions the programmer had to "speak" machine language. This language was and is usually hexadecimal (base 16), octal (base 8), or binary (base 2). You do not have to know the details of hexadecimal — often called hex — or octal. You only need to know that the terms exist and refer to machine language. Machine language is a series of numbers. Using machine language is similar to using Morse code to communicate with another person. If someone was unable for some reason to use language and only "understood" a sequence of dots and dashes, then Morse code would be the mechanism for communication. A record of such a conversation would be a series of dots and dashes. A printout of machine language programs is often a series of numbers. There are people who have worked with this code for enough time to be able to "read" this record easily. There are people who have worked with Morse code for enough time to be able to read pages of Morse code conversation easily. Clearly if we all had to use Morse code to talk to each other it would be slow, frustrating, error prone, and perhaps boring. Using machine language to give instructions to a computer is time-consuming, frustrating, and at times boring.

To avoid errors, people developed higher level languages containing those higher level commands that we spoke of earlier. The development of these higher level commands is similar to the development of a Morse code machine that can take a typed word and translate it into Morse code for sending. The machine should also be able to take a Morse code signal and translate it into a typed word. Through this machine a non-Morse code literate person could communicate with a Morse code only person. All you would have to do is use this machine and type in words that the machine could translate into Morse code. You would have to know the commands that the Morse code machine could use, but you would not have to know Morse code. If the Morse code translation machine can create the Morse code more quickly than the human being, then use of this machine can speed up communication via Morse code. There are many higher level languages. Some are more powerful than others. Many modern languages have low-level operators such as "+" and "-" that work essentially like their low-level language counterparts. The modern languages make things easier (handle complexity) or safer (make sure that types match).

The input (code) that the machine understands (the Morse code) is called object code. The code that a human understands is called source code. Object code is fit for machine consumption, and source code is fit for human consumption. A program called a compiler turns source code into object code. Compilers are translators that translate source code into object code. If you display source code on your computer screen, it looks like English (you may not understand the grammar or the meaning, but it looks like English). If you try to display object code, either nothing or garbage will print. Most of the object code characters have no human-readable character representation. To modify a program you need the source code. Why? So you know what is going on with the program. Almost all software that you buy, however, contains only the object code. And in some industry circles the issue of whether to release source code is a hot one.

In the past, the source code was usually released with the object code so both you and your machine could understand what was going on. When the source code is released, you can make changes to the software. This is handy if you need modifications because you can then make the modifications yourself (if you know the programming language). This is handy for the programmer(s) who developed the application because they do not have to spend time modifying the program. If there is a bug in the program, you might also be able to fix it. This benefits both you and the programmer(s) who developed the software. You get the bug fixed quickly and the programmer(s) have many people helping them debug the program. There are disadvantages for the application developers when the source code is released. Everybody knows exactly how the program works. Everybody knows their "secret ingredients" and trade secrets. Giving the source code away often means giving years of work and creativity away for free for others to copy. When others copy all or part of the source code they usually do so to include it in their products. Giving away the source code means that you are doing development work for yourself and your competitors. Source code was given away in the days when computing was noncompetitive. In those days people gave the source code away largely because they were happy that someone wanted to use their code and because they were happy to have others help them debug it.

Programming languages differ in their utility. That is, some programming languages are considered to be better for some applications than others. For example, Cobol is generally considered to be a business language. Fortran is considered to be a scientific language. Lisp is used in artificial intelligence work. Some are considered to be more sophisticated than others. The languages that you know can determine in which areas you work. The languages that you know determine your employability especially if you only know a language implemented on a few machines or used by a few people.

Mass-Produced Applications Packages

Again, a programming language is used to write application software. An application package is what is commonly bought for use in the workplace. A mass-produced applications package, is mass produced for the mass market. Such a package is often called "off-the-shelf" software. When you use workplace automation you do not usually want to learn a programming language, you just want to be able to do the things required in your job. Programmers develop applications that allow you to do this without you having to know a programming language. The instructions or commands of the applications program are part of the user interface. The programmer instructs the computer what to do with each of the commands that you issue. Packages differ in the commands that compose the user interface. Commands used in applications packages are also not the same from package to package. Each time you use a new applications package you have to learn the commands for that package. The command to print a document in one package, for example, may be hitting the "P" key, whereas in another it might be highlighting the word "Print" displayed on a menu. The intent with an interface is to have the user issue commands that are easy for the user to remember and, when forgotten, easy to reconstruct. Some packages accomplish this and some do not. Also the commands and method of using a package vary from package to package usually depending on the developer.

Customized Software

Sometimes the application required for an organization is not commercially available — it has not been written. In this case the company has three choices:

1. Hire programmers to write the application in-house.

2. Hire consultants to write the application.

3. Forget the whole thing and wait until something the company can use is developed.

The first two choices mean that the company has decided to enter the customized application package process. This process should not be undertaken lightly. The development of a customized application package is similar to the development of a customized house. Development of a customized application requires successful completion of two major tasks:

1. Development of the system specifications.

2. Development of the system software according to the specifications.

The process of developing system specifications is analogous to developing the blueprints for a house. Detailed blueprints for a house are developed through the interaction of a customer and the architect. The architect turns the general ideas and needs of the customer into the detailed plans required by the builder. Detailed computer system specifications are developed through the interaction of the potential users and an information specialist. The information specialist turns the general ideas and needs of the potential user into the detailed plans required by the programmer. The development of system software is analogous to the construction of a house. The programmer is similar to the builder. The machine with which the programmer works is similar to the building supplies used by the builder. Customized software is not cheap. It may take five or six programmers a year to do even a poor word processing package from scratch. You might end up paying a couple of hundred thousand dollars for a customized package not much better (or even worse) than a mass-produced package you could have bought for a couple of hundred dollars.

Chapter 12
Technology Transfer
Summary

There were plenty of ways to get rid of that officer by some simple and plausible device, but no, I must pick out a picturesque one; it is the crying defect of my character.

Mark Twain in *The Connecticut Yankee in King Arthur's Court,* Chapter XXXVII, An Awful Predicament.

"Ruin, Ludlow Castle, Shropshire, England," photo © copyright Christi Carter, 1993.

There is nothing picturesque about this final summary chapter. It is a simple chapter that lists all the questions that must be answered for successful technology transfer to occur. A few guidelines are also mentioned. Answering these questions provides a path through and a doorway out of the confusion of technology transfer. There is nothing picturesque about the path, but that does not mean it is not elegant.

Six Technology Transfer Decision Stages

Figure 12.1 is a schematic view of the technology transfer process. For workplace computing technology to transfer successfully to a business customer, the customer must decide and choose:

1. Which organizational problem (if any) needs solving

2. Which organizational resources (if any) are available

3. Which general solutions (if any) are possible

4. Which solution technologies (if any) are possible

5. Which solution products (if any) are possible

6. Which methods of product use are effective/optimal.

Stage One Questions: Realize the Problem

Successful completion of Stage One (decide and choose which organizational problem, if any, needs solving) requires that the following seven questions be answered:

1. What is the company's mission?

2. What are the company's goals and objectives?

3. What priority is each goal and objective?

4. How effectively is each goal and objective being met?

5. Which ineffectively met goal or objective, if any, needs to be addressed immediately?

6. What are the problems of meeting this goal or objective at each level of the organization?

7. What problem, if any, needs to be addressed and solved immediately?

Figure 12.1 Six Stage Technology Transfer Process

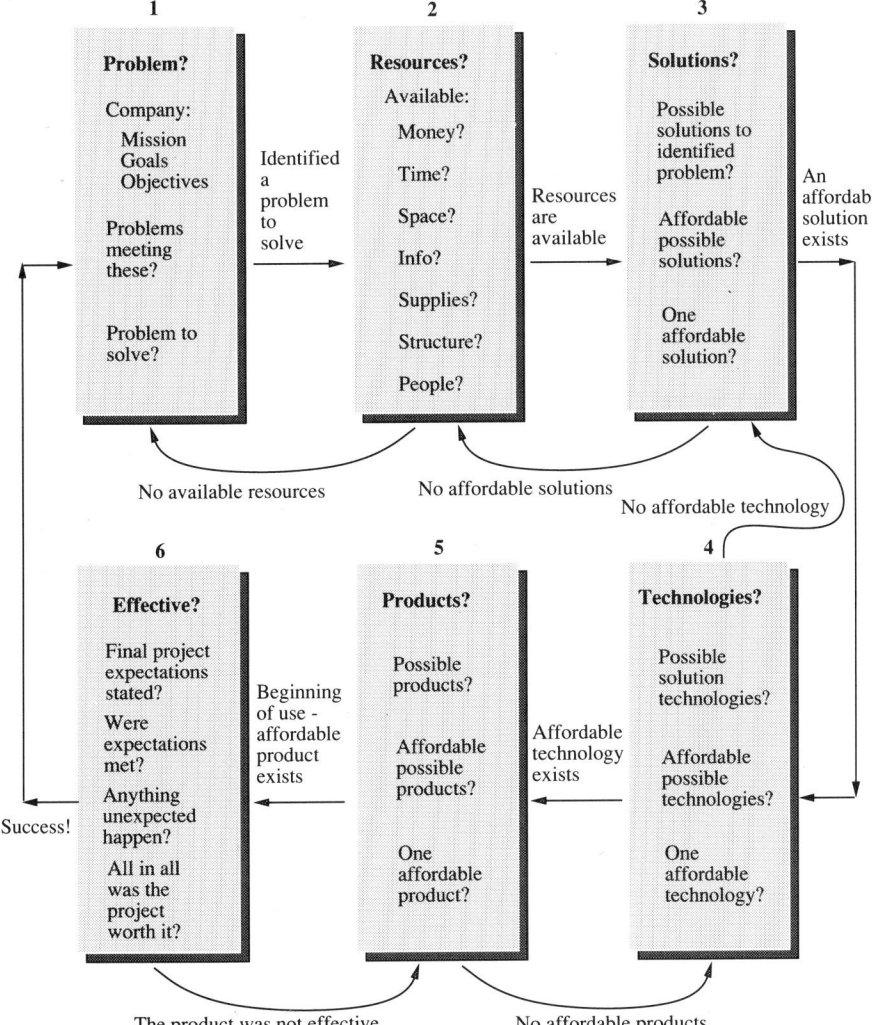

Answering Question Six (problems of meeting goal) requires answering four questions:

1. Who are the relevant people for Question 6?

2. What are the stated/presented problems to goal achievement?

3.   What are the problems for discussion?

4.   What are the identified problems?

Stage Two Questions: Thriving Resources

Successful completion of Stage Two (decide and choose which organizational resources, if any, are available) requires that the following eight questions be answered:

1.   Who are the relevant people for this stage?

2.   How much money is available for the project?

3.   How much time is available for the project?

4.   How much space is available for the project?

5.   What information is available for the project?

6.   What supplies are available for the project?

7.   What structure is available for the project?

8.   Which people are available for the project?

To determine how much money is available, you must answer the following three questions:

1.   Is there a cap? Is there some amount of money above which you do not want the cost of the project to go? (What are the upper and lower bounds of using available money?)

2.   How much money is available over a short period of time? (How much money is available today from a bank account or a loan?)

3.   How much money is available over a long period of time? (What is the cash flow over time available for the project?)

To determine how much time is available, you must answer the following three questions:

1.   Is there a deadline? (Is there some point in time by which the project must be finished?)

2.   How much time is available on a routine daily basis?

3.   How much time is available to handle emergencies/crises?

Stage Three Questions: Schemes and Solutions

Successful completion of Stage Three (decide and choose which general solutions, if any, are possible) requires that the following 13 questions be answered:

1. Who are the relevant people for this stage?

2. What are the stated/presented general solutions to this problem?

3. Did Question 2 provide any information that warrants reevaluation of prior decisions?

4. What are the general solutions for discussion?

5. What are the identified solutions for the problem?

6. Did Question 5 provide any information that warrants reevaluation of prior decisions?

7. What is the estimated cost of each identified solution?

8. Did Question 7 provide any information that warrants reevaluation of prior decisions?

9. Which solutions, if any, are affordable?

10. Did Question 9 provide any information that warrants reevaluation of prior decisions?

11. Which solutions, if any, will be chosen for implementation at this time?

12. What is the budget and schedule for the rest of the project?

13. Did Question 12 provide any information that warrants reevaluation of prior decisions?

To generate "ballpark" costs (Question 7), you need to answer the following seven questions:

1. How much money is required to complete the project successfully?

2. How much time is required to complete the project successfully (e.g., 100 person-hours, 3000 person-hours)?

3. How much and what type of space is required to complete the project successfully (e.g., 200 square feet of storage space, raised floor, and climate-controlled room)?

4. What information is required to complete the project successfully (e.g., current functioning information, difference between computers)?

5. How much and what type of supplies are required to complete the project successfully (e.g., 200 pens, 150 pads of paper, 2 computers, network cables, modems, 8 file cabinets, 2 tables for work)?

6. What structure is required to complete the project successfully (e.g., unrestricted flow of information between those involved in the project, self-discipline, motivation)?

7. How many and which people are required to complete the project successfully (e.g., all members of the accounting department, temporary employees)?

The budget and schedule should be listed in the project notebook. Although different companies may include different pieces of information in their notebook, we recommend that at least the following information be included:

1. Original budget

2. Budget revisions

3. Current working budget

4. Original schedule

5. Schedule revisions

6. Current working schedule

7. List of "Things to Do" to meet schedule

8. List of people working on the project

9. List of people responsible for project tasks and activities

10. Information used in decision-making (e.g., product information)

11. List of key decisions and reasoning for each decision

12. List of problems encountered and solutions

13. General project progress notes (can include minutes from meetings)

14. Inventory of purchased products.

Stage Four Questions: Technologies Insolent and Grand

Successful completion of Stage Four (decide and choose which solution technologies, if any, are possible) requires that the following 15 questions be answered:

1. Who are the relevant people for this stage?

2. What are the specific tasks (problems to be solved) of the identified solution?

3. Did Question 2 provide any information that warrants reevaluation of prior decisions?

4. What technologies might be used to complete these specific tasks (solve these specific identified solution problems)?

5. Did Question 4 provide any information that warrants reevaluation of prior decisions?

6. What are the technologies for discussion?

7. What are the identified technologies?

8. Did Question 7 provide any information that warrants reevaluation of prior decisions?

9. What is the cost of each identified technology?

10. Did Question 9 provide any information that warrants reevaluation of prior decisions?

11. Which technologies, if any, are affordable?

12. Did Question 11 provide any information that warrants reevaluation of prior decisions?

13. Which technology, if any, will be chosen for use at this time?

14. What is the budget and schedule for the rest of the project?

15. Did Question 14 provide any information that warrants reevaluation of prior decisions?

Stage Five Questions: Products Plain and Fancy

Successful completion of Stage Five (decide and choose which solution products, if any, are possible) requires that the following 26 questions be answered:

1. Who are the relevant people for this stage?

2. What are the specific tasks expected of chosen products?

3. Did Question 2 provide any information that warrants reevaluation of prior decisions?

4. What products are available that can complete these specific tasks?

5. Did Question 4 provide any information that warrants reevaluation of prior decisions?

6. What are the products for discussion?

7. What are the identified products?

8. Did Question 7 provide any information that warrants reevaluation of prior decisions?

9. What is the cost of each identified product?

10. Did Question 9 provide any information that warrants reevaluation of prior decisions?

11. Which products, if any, are affordable?

12. Did Question 11 provide any information that warrants reevaluation of prior decisions?

13. Which products, if any, will be purchased at this time?

14. What are the available product vendors?

15. Did Question 14 provide any information that warrants reevaluation of prior decisions?

16. What are the similarities and differences between vendors?

17. Did Question 16 provide any information that warrants reevaluation of prior decisions?

18. Who are the vendors for discussion?

19. Who are the identified vendors?

20. Did Question 19 provide any information that warrants reevaluation of prior decisions?

21. Who will do product installation?

22. Did Question 21 provide any information that warrants reevaluation of prior decisions?

23. How will initial training be completed?

24. Did Question 23 provide any information that warrants reevaluation of prior decisions?

25. What is the budget and schedule for the rest of the project?

26. Did Question 25 provide any information that warrants reevaluation of prior decisions?

The following purchase guidelines should be followed when purchasing computers and computer-related products:

1. Choose the product of greatest utility. (For computers, this means choosing the software before choosing the hardware.)

2. Try to purchase a computer with the greatest compatibility. Pay careful attention to the computer's processor.

3. Include both short-term and long-term costs in your calculation of total product cost.

4. Look for the hidden costs (e.g., inexpensive hardware and expensive software, inexpensive software with expensive hardware, high maintenance cost).

5. Consider the resale price of purchased software and hardware.

6. Purchase only reputable, high-volume products. Such products are usually of the best quality at the lowest cost.

7. Purchase a fairly new product.

8. Consider the full range of product alternatives — even ones that seem extreme — because this will clarify your automation picture for you. Clarification may reveal a low-cost solution that is as effective as a higher priced solution.

9. Purchase one type of machine.

10. Do not buy more than you need.

11. Try to buy in quantity or bulk since this usually reduces the cost of any individual item.

Stage Six Questions: Let the Facts Reign

Successful completion of Stage Six (decide and choose which methods of product use are effective/optimal) requires that the following three questions be answered:

1. Who are the relevant people for this stage?

2. When, exactly, will the project end?

3. How effective/successful was the project?

To answer the second question, you need to determine three different project end points:

1. When does the initial adjustment period end?

2. When does the evaluation period end?

3. When is the final report on project success due?

To determine project success you need to answer five questions:

1. What did you expect the project to accomplish?

2. How will you measure these expectations?

3. To what extent were these expectations met?

4. Did anything unexpected happen?

5. All things considered, was the project worth the effort?

Outside Consultant Guidelines

1. Automation projects do not always require the use of an outside consultant.

2. If the company decides to use an outside consultant, choose one the same way any other professional consultant is chosen. Automation consultant selection issues are the same as selection issues for consultants in any other area.

3. Do not relinquish decision responsibility and control to the consultant.

4. If the company uses an outside consultant for any part of the automation project, establish the total project price, project specifications, completion dates, and deadline details *first*.

5. Avoid packages of odd hardware and software that only the consultant can operate/repair/distribute. The automation project should not produce future revenues for the consultant.

6. Avoid outside consultants who also sell hardware and software package deals. Also avoid high-cost "training" packages.

7. Avoid outside consultants with special or narrow organizational interests and expertise. Choose consultants who consider the organization as a whole and not just one aspect.

"Bacchante in Lime Walk, Sissinghurst Garden, Kent, England," photo © copyright Christi Carter, 1993.

The End

Index

D

E

F